Time Travel

Time Travel

The Popular Philosophy of Narrative

David Wittenberg

FORDHAM UNIVERSITY PRESS

NEW YORK 2013

Fordham University Press has no responsibility for the persis-
tence or accuracy of URLs for external or third-party Internet
websites referred to in this publication and does not guarantee
that any content on such websites is, or will remain, accurate or
appropriate.

Fordham University Press also publishes its books in a variety of
electronic formats. Some content that appears in print may not
be available in electronic books.

Library of Congress Cataloging-in-Publication Data

Wittenberg, David.
   Time travel : the popular philosophy of narrative / David
Wittenberg. — 1st ed.
        p. cm.
   Includes bibliographical references and index.
   ISBN 978-0-8232-4996-1 (cloth : alk. paper)—
ISBN 978-0-8232-4997-8 (pbk. : alk. paper)
     1. Time perception in literature.   2. Time travel in
literature.   3. Narration (Rhetoric)   4. Literature—
Philosophy.   I. Title.
   PN56.T5W58 2013
   809'.93384—dc23

                                        2012029101

15 14 13   5 4 3 2 1
First edition

for Lara and Leela

CONTENTS

ACKNOWLEDGMENTS

A great number of people helped me think through and complete this book. Thanks to Holly Carver, Dilip Gaonkar, Paul Harris, Brooks Landon, Rob Latham, Tom Lay, Tom Lutz, Julia Madsen, David Martinez, Eric Newman, Joe Parsons, Lily Robert-Foley, Garrett Stewart, Satomi Saito, Michael Witmore, Agnes and Albert Wittenberg, Haihong Yang, and my undergraduate and graduate students at the University of Iowa. Thanks also to my colleagues in the International Society for the Study of Time, and to the curators and librarians at the superb Eaton Collection of Science Fiction and Fantasy at the University of California, Riverside. Special thanks, as always, to Judith Butler, Jean McGarry, and Helen Tartar. Most of all, thanks to Lara Trubowitz, whose support is vital to me in every way.

I gratefully acknowledge Byron Preiss Visual Publications for permission to reprint John Jude Palencar's illustrations from the 1981 *Distant Stars* version of Samuel Delany's *Empire Star* (illustrations © Byron Preiss Visual Publications). An earlier version of my Chapter 6 appeared in the journal *Discourse* with the title "Oedipus Multiplex, or, The Subject as a Time Travel Film: Two Readings of *Back to the Future*"; thanks to Wayne State University Press for permission to reprint this material. I am grateful to Cary and Michael Huang for permission to reproduce a screen capture of their interactive web tool, "Scale of the Universe."

Time Travel

Introduction: Time Travel and the Mechanics of Narrative

> "What happened to me?" I whispered to the lady at my side.
>
> "Pardon? Oh, a meteor got you, but you didn't miss a thing, believe me, that duet was absolutely awful. Of course it *was* scandalous: they had to send all the way to Galax for your spare," whispered the pleasant Ardritess.
>
> "What spare?" I asked, suddenly feeling numb.
>
> "Why, yours, of course."
>
> "Then where am I?"
>
> "Where? Here in the theater. Are you all right?"
>
> "Then I am the spare?"
>
> "Certainly."
>
> —STANISLAW LEM, *The Star Diaries*

Anyone who thinks about time travel for a while is likely to encounter something like the following dilemma. On the one hand, time travel stories would seem to constitute a minor and idiosyncratic literature, a subtype of other popular genres such as science fiction, romance, and action-adventure; time travel makes use of improbable devices and extravagant paradoxes, and in general lays claim to only a small share of the plots, topics, or themes that could conceivably interest a reader, writer, or critic of literature. On the other hand, since even the most elementary narratives, whether fictional or nonfictional, set out to modify or manipulate the order, duration, and significance of events in time—that is, since all narratives do something like "travel" through time or construct "alternate" worlds—one could arguably call narrative itself a "time machine," which is to say, a mechanism for revising the arrangements of stories and histories. In this more expansive view, literature itself might be viewed as a subtype of time travel, rather than the other way around, and time traveling might be considered a fundamental condition of storytelling itself, even its very essence.

This book sets out, not exactly to resolve such a dilemma over the significance of time travel stories, but rather to amplify and further complicate it, even to expose some of its more provocative implications. In this respect, the book is polemical as well as analytical. I contend that there ought to be much more attention paid to the seemingly eccentric genre of time travel fiction by literary, cultural, and film theorists, as well as by readers and scholars interested more broadly in either theories of narrative or philosophies of time. If it eventually turns out that the question of time travel is tantamount to the question of narrative itself, then such a fundamental question will almost certainly have been woefully underconsidered by the very thinkers best positioned to comprehend and answer it.

I argue that time travel fiction is a "narratological laboratory," in which many of the most basic theoretical questions about storytelling, and by extension about the philosophy of temporality, history, and subjectivity, are represented in the form of literal devices and plots, at once both convenient for criticism and fruitfully complex. I wish to suggest not merely that time travel stories are examples or depictions of narratological or philosophical issues, but that these stories are themselves already exercises in narratology and the theorization of temporality—they are in essence "narrative machines," more or less latent, emergent, or full blown. And following the leads they expose, the present book intends to contribute its part to a fundamental reconsideration of the philosophy of time, as well as to a fundamental synthesis of such philosophy with narrative theory, goals in the service of which time travel fiction will be regarded as a philosophical literature *par excellence*.

If it seems brash for an academic critic to make such broad philosophical claims for a popular genre and, moreover, for a genre that has previously garnered only sporadic attention from academic criticism, the brashness may be mitigated by the book's inevitably interdisciplinary focus. By necessity, I borrow and amalgamate a range of insights and information from cultural and film theory, philosophy (both analytic and continental), physics, psychology, and historiography. In each case, interpretations of the literature of time travel, chiefly of popular science fiction, serve as a kind of escort into and around these other fields. Let me begin with a few such interpretations. Each of the following three readings analyzes a single time travel story as a paradigm of a certain form of temporal manipulation, and in turn as an access into crucial aspects of narrative theory. Following these three readings, I briefly outline some of the conceptual and methodological links that such interpretations might suggest between fields of study concerned with problems of time travel—in particular, popular literature and film, literary history and criticism, analytic philosophy, and physics. I also comment on the somewhat counterintuitive history of the genre of time travel fiction that I necessarily construct alongside the specific theoretical considerations time travel stories compel me to pursue.

First Reading: Fabula *and* Sjuzhet *in* Up the Line

In Robert Silverberg's 1969 novel *Up the Line*, the following conversation takes place between Jud Elliott, a transtemporal tour guide for researchers and vacationers journeying into the past, and his mentor in the "Time Service," Themistoklis Metaxas. In a side plot, Metaxas is helping Jud get invited to a party in Constantinople in the year 1105, so that Jud can seduce his own "great-great-multi-great-grandmother":

> Metaxas, as always, was glad to help.
> "It'll take a few days," he said. "Communications are slow here. Messengers going back and forth."
> "Should I wait here?"
> "Why bother?" Metaxas asked. "You've got a timer. Jump down three days, and maybe by then everything will be arranged."
> I jumped down three days. Metaxas said, "Everything is arranged."[1]

Here is a seemingly extraordinary narrative event, one peculiar to the temporal manipulations made possible by a time machine: the narrator, Jud, using his "timer," is able instantaneously to skip over three full days of time, meeting up with Metaxas in the same location "down the line."[2] Silverberg constructs the conversation to reflect the unusual temporal elision enabled by the timer, skipping instantaneously between the two disconnected fragments of dialogue and treating them as continuous: "I jumped down three days . . . '[e]verything is arranged.'" The reader's own perspective, which has also been made to "jump" over the same three days, is therefore allied with Jud's experience of the hiatus rather than, for instance, with that of

Metaxas, who has presumably been occupied during the entire interval. In turn, Silverberg can joke, via the deadpan substitution of tenses in the phrase "everything will be arranged/is arranged," about the economic convenience the time machine provides for Jud, who now proceeds directly to his tryst, bypassing the labor of preparations undertaken by his colleague Metaxas. For a brief moment, whatever other unusual advantages it may offer, time travel permits the indulgence of an erotic perquisite, one that the reader shares by being positioned all the nearer, in terms of the economy of reading, to the imminent seduction.

Although the most extraordinary element in this fragment of dialogue is surely the presence of the time machine (the timer), it is not easy to determine precisely in what its extraordinariness consists. Consider this slight rewriting of the scene, in which I merely replace the time traveler's technical jargon with some more mundane language:

> Metaxas, as always, was glad to help.
> "It'll take a few days," he said. "Communications are slow here. Messengers going back and forth."
> "Should I wait here?"
> "Why bother?" Metaxas asked. "You've got things to do. Come back in three days, and maybe by then everything will be arranged."
> I came back in three days. Metaxas said, "Everything is arranged."

With all explicit reference to timers and jumping removed, we now have a perfectly ordinary sequence of narrative events, in which the narrator-protagonist also skips, still instantaneously from the reader's viewpoint, three days of time. The vague "things to do" with which I have replaced "a timer" in my new version, and that Metaxas now suggests might account for Jud's missing three days, are, in terms of the progress of the narrative, exactly as nondescript and formally empty as the three days originally "jumped" using time travel. Even the humor entailed by Jud's avoidance of the labor Metaxas has undertaken on his behalf can be retained in the deadpan echo of the final line. Indeed, whether the skipping of three days consists, for Jud, in a physically discontinuous *nothing* enabled by a time machine or whether, instead, he merely expends three days in the background of the narrative doing nothing much, makes very little difference to the structure or coherence of the fiction, even though it may, of course, make considerable difference to the story's genre.

In short, physical time travel and metanarrative juxtaposition are, in narratological if not in generic terms, identical. Whether such identity is an artifact of the way in which specific stories are constructed or whether it has some more profound and wide-ranging narratological or even ontological significance—whether, in other words, the metanarration of time travel is truly a basic feature of the way in which we tell even the most conventional stories—remains an open question, and one that I will pursue in greater detail especially in the book's second half. For now, the apparent structural equivalence of the time travel plot with more conventional plots should at least indicate how a time machine might duplicate some of the fundamental actions of narratives generally. The timer appears to do exactly what plots do already, but in some sense more *literally*.

As Mieke Bal notes, within conventional narratives, temporal discontinuities, dilations, and repetitions occur constantly, "often without being noticed by the reader."[3] Indeed, in most stories, quite drastic manipulations of chronology on the level of form—hiatuses, flashbacks, sudden temporal cuts, overlapping events—are cheerfully tolerated by the story's audience. Even when such manipulations are directly foregrounded, for instance by an explicitly reminiscing chronicler or by a Scheherazadean metanarrator, the reader usually has little difficulty receiving such plotlike narratological exposures as unobtrusive (if that is the author's goal) and, in a word, "normal," or as what Gérard Genette calls "classical."[4]

However, in a time travel story, even the most elementary experience of plot involves an essentially abnormal metanarrative intervention, since the "classical" mechanisms of temporal discontinuity, dilation, or reordering are now introduced directly into the story itself, in the guise of literal devices or mechanisms. They are no longer either tacit or formalistic but rather actual and eventlike—or, in terms of the fiction itself, *real*—a fact that makes time travel fiction already, and inherently, a fiction explicitly about the temporality of literary form. Simply to follow the action of *Up the Line*, the reader must directly relate the two divergent time frames experienced by Jud and Metaxas, and then further compare them against the hypothetical background of a metaframe that leaves entirely open the question of whether the original two time frames might ultimately be reconciled. Indeed, it is this potential irreconcilability—the possibility that, for instance, Metaxas could kill rather than solicit Jud's "great-great-multi-great-grandmother"—that gives the science fiction reader, as well as the consumer of time travel stories in the broader culture, a glimpse of what is known as a paradox story. But even where it doesn't eventuate in paradox, or for that matter in any logical or causal conundrum, the potential irreconcilability of narrative frames within a time travel story still potentially imparts to the reader's experience an unusual and subversive novelty, one that automatically exposes and destabilizes some of the basic conditions of story construction.

In short, the novelty of the time machine is simultaneously *outré* and utterly basic to what we accept as normal or classical storytelling. We are able to see this most clearly in narratological terms, because the time travel plot is essentially a creative abuse of the usual narratological rules, or even a direct mimesis or parody of narrative formation. Narratologists, following the tradition of early formalist criticism, distinguish between "story" and "plot," or what the Russian formalists termed *fabula* and *sjuzhet*, a terminology I will revisit at greater length in Chapter 4.[5] *Fabula* is the ostensible underlying sequence of story events in a narrative, *sjuzhet* its re-formation as a specific plot, the reconstructed montage of story elements arranged by an author within a given set of generic rules or protocols. In conventional narratives, temporal alterations such as changes of order or pace, repetitions, and the skipping of time occur on the level of *sjuzhet*, and the rules governing plot construction within genres allow for considerable and even radical variations in the order or frequency of *sjuzhet* events. So when Clarissa Dalloway is suddenly no longer on Bond Street but rather back at her house, or when we see Charles Foster Kane decrepit and dying in an early scene of *Citizen Kane* and then as a young child in a later scene, such chronological anomalies or "anachronies"[6] are immediately understood to be artifacts of plot manipulation arranged on the level of *sjuzhet*, and not characteristic of the underlying *fabula*, which is presumed to remain linear and chronological. Such a presumption is tantamount to asserting that neither *Mrs. Dalloway* nor *Citizen Kane* is either a supernatural or a time travel story, and that we (readers, writers, audience, and critics) continue to assume that Clarissa "really" takes more or less the usual number of minutes to walk home from Bond Street, and that Charles Foster Kane "really" grows from childhood to old age instead of the reverse. Indeed, the presumption of chronological regularity within the underlying story material is crucial for the coherence of the classical narrative. Otherwise, the potentially extreme temporal variations of the *sjuzhet* would emerge as merely fragmented and anachronic. Indeed, they can easily be made to emerge that way within a variety of experimental fictions, for instance in Jorge Luis Borges's "Garden of Forking Paths," Martin Amis's *Time's Arrow*, or Philip K. Dick's *Time Out of Joint*. We might even preliminarily define the "normal" narrative—so closely related to Genette's "classical"—as one in which divergences from regular chronology occur *only* on the level of the *sjuzhet*, never in the *fabula*.[7]

By contrast, in a time travel fiction, even a relatively normal one, no such underlying coherence in the *fabula* may be assumed. A time machine potentially alters the chronology of story events themselves, making it impossible to presuppose or determine any single consistent relationship between *fabula* and *sjuzhet*, and requiring, therefore, more or less artificial or narratively supplemental mechanisms of coherence. In Silverberg's *Up the Line*, the interval over which Jud jumps using the time machine cannot, in principle, be recovered or reconstructed in the mode of three days that really passed in the *fabula* while only the reader jumped over them in the *sjuzhet*. Here it is no longer possible at all to decide whether these three days have really taken place without first selecting a privileged viewpoint other than Jud's (for instance, that of Metaxas), or without explicitly adopting some further metanarrative frame within which the several actual or potential views of the story's chronology might be contrasted and adjudicated. It would not even be possible to declare that Jud's three days are "gone," excised altogether from the *fabula*, since any time travel story necessarily offers multiple interpretations of the existence or nonexistence of such durations and of the agents who experience or witness them. We must say instead that the ontological status of these three days, as well as their partial or quasi assimilation by the domains of either *fabula* or *sjuzhet*, has been rendered radically ambiguous. Such radical structural ambiguity becomes even more obvious as soon as time travelers begin to do more unusual things, such as relive their own pasts, meet or duplicate themselves, retroactively eliminate slices of history, reexperience those same slices in altered versions or "lines," and so on. What "normal" narratives can bring about only in the form of fantasy, allegory, or formalistic experimentation, time travel narratives accomplish in the mode of unfussy realism, a literal or mimetic description of characters, events, and machines. Hence "anachrony" in a time travel story can never be either dismissed as provisional or finessed as a mere artifact of retelling; it belongs ineluctably to the *fabula* itself, and remains fully present in all its potentially paradoxical provocation.

In this sense, time travel fiction directly represents, on the level of straightforward content, not only the processes by which narratives are formed, but also the experimental conditions under which controlled narratological inquiry might take place, and "normal" or "classical" narrative procedures and techniques be manipulated and productively malformed. Even the naïve reader or audience of a time travel fiction becomes, by default or exigency, a practicing narrative theorist or a practical experimenter in the philosophy of time. As I show in more detail beginning in Chapter 3, all narratives, even "normal" ones, can be read only in a kind of hyperspace or metaverse, a quasi-transcendental "space" in which the relation between always potentially divergent lines of narration is negotiated and (usually) brought back to coherence and synchrony. The reader of time travel fiction, even in its most mainstream or adolescent-literary modes, is entirely familiar with this hyperspatial or metaversal realm of narratological negotiation. It is the very medium through which the time-traveling protagonist, who is *de facto* never either fully in or fully outside of the plot, realistically travels. Narratology is the very *mise-en-scène* of time travel fiction, and time travel itself the machinery by which narrative is manufactured.

Second Reading: Psychohistoriography in Behold the Man

In my second reading, from Michael Moorcock's 1968 novel *Behold the Man*, the sorts of eccentric plot twists encountered in Silverberg's humorously narcissistic and oedipal adventure are considerably dampened, even as *Behold the Man* contrives to maintain in more polemical tension much of the potentially paradoxical relation between its *fabula* and *sjuzhet*. Indeed, Moorcock inflects this relation with a pathos engendered by the difficulty of interpreting and retelling, with any real precision and consistency, our most elemental or influential stories and histories. Overall, Moorcock is considerably more interested in the underlying psychological motivations of such temporal and historical manipulation, which for Silverberg remain largely a formalistic game.

Moorcock's protagonist, Karl Glogauer, an alienated and melancholic late-modern subject, exhibits an unhealthy obsession with two types of historical questions: first, his own recurring self-destructive relationships with women, and, second, the apparently more objective or pedantic theological question of the divinity of Christ. Being keen to circumvent the former question by way of pursuing the latter, and having become acquainted with the inventor of a time machine, Glogauer offers himself as test subject for a time travel experiment, on condition that he be sent back to A.D. 29 in order to witness the crucifixion of Jesus. Following a crash landing in the desert outside Jerusalem, in which Glogauer's delicate, liquid-filled time machine bursts open and he is stranded in the past, he soon acclimates very nicely to the first century, even given his limited knowledge of the local dialect of Aramaic. He gradually adopts the superstitious and self-mortifying culture of John the Baptist and his sect of Essenes. However, despite a determined search, Glogauer has difficulty turning up any evidence of Jesus of Nazareth, or even of Mary or Joseph. When he finally does track down the "holy family," he discovers that Joseph has deserted his wife, Mary herself is anything but virginal, and, to his greatest disappointment, the adolescent Jesus is misshapen and mentally disabled, an unpromising candidate for a messiah.

However, by the time Glogauer gets around to making these unsettling discoveries, his odd and repeated inquiries about a Nazarene carpenter named Jesus, along with his own anachronistic quirks of character and his apparent ability to predict impending events with uncanny detail, have attracted attention in Nazareth and Bethlehem. The locals begin to consider him a prophet.[8] As the novel proceeds toward its conclusion, Glogauer embraces the prophetic role into which his incongruity has thrust him, and willingly acts the part of Jesus of Nazareth. He then sets about to orchestrate the series of events that he knows, with an ironically detailed retrospect, to have made up the end of Jesus's life. Ultimately, he arranges to have himself crucified by the Romans, all the while scrupulously duplicating Jesus's specific gestures and expressions, which he recalls from the Gospels and histories he had obsessively studied in his "earlier" life two millennia hence.

Unlike Silverberg, Moorcock is not concerned with creating suspense about the paradoxical twists in his plot. Relatively early in the novel, it becomes clear what is going to happen to Karl Glogauer, although, in proportion to the strength of his own delusions, Glogauer himself is slower than the reader to figure it out. Here, for instance, is his initial conversation with John the Baptist just after crashing in the desert and being rescued by the Essenes:

> "So you are from Egypt. That is what we thought. And evidently you are a magus, with your strange clothes and your chariot of iron drawn by spirits. Good. Your name is Jesus, I am told, and you are the Nazarene."
>
> The other man must have mistaken Glogauer's inquiry as a statement of his own name. He smiled and shook his head.
>
> "I seek Jesus, the Nazarene," he said.[9]

What Glogauer seeks, however, is both the actual man Jesus and, in a sense that is not yet apparent to him, a more complex amalgam of theological and psychological cathexes, which in a more straightforwardly modern theological vocabulary might be described as the "sense of Christ within" him, or, in a psychoanalytic vocabulary, as an ego-ideal. In short, Glogauer himself now gradually evolves into the person or archetype he seeks, a halting and uncanny change for him because he remains unaware of the full range of his own motivations:

> Karl Glogauer grew his hair long and let his beard come unchecked. His face and body were soon burned dark by the sun. He mortified his flesh and starved himself and chanted his prayers beneath the sun, as [the Essenes] did. But he rarely heard God and only once thought he saw an archangel with wings of fire.
>
> One day they took him to the river and baptized him with the name he had first given John the Baptist. They called him Emmanuel.
>
> The ceremony, with its chanting and its swaying, was very heady and left him completely euphoric and happier than he ever remembered.[10]

In *Behold the Man*, the mechanics of time travel, now in the hands of a more contemplative and psychologically incisive writer than Silverberg, readily permits complex intersections between a number of the book's themes: Glogauer's neurotic obsessions with certain sexual objects (for instance, crucifixes between women's breasts); Glogauer's half-hearted efforts to do away with himself, his "martyr complex"; Glogauer's obsession with the theological problem of Christ's divinity; Glogauer's sense of dislocation or alienation in the present day, his feeling that he would be more at home in the past; the philosophical or psychological significance of Jesus as a paragon of self-sacrifice or self-mortification; the variety of sexual subtexts of Christian doctrine; the question of the relation between Christ's godliness and an individual person's inherent divinity; the psychical and sociological motivations of the actual historical Jesus and his followers; and the significance of fate (e.g., in Jesus's prediction of Judas's betrayal) and its ironic correlation with the "predictions" of the time traveler. Of course, with the exception only of the last one, these are all themes raised in any number of other literary, philosophical, and theological works. However, a time travel narrative, because of the way in which its literal devices bear upon or even directly affect the basic structure of the narrative itself, may be more likely than most popular literary forms to raise several such issues simultaneously. In that sense, presumably, the value of a book such as *Behold the Man* might be its capacity, within a relatively accessible generic medium, to address problems that are usually relegated to "higher" literary types, or at least to more self-consciously experimental and philosophically oriented ones.

But the more substantial difference that the time travel story makes, as I also wished to show in the Silverberg example, is not the specific theoretical or philosophical issue at hand, nor its ostensible or unusual level of complexity, but rather the *mode* in which that issue is woven into the substance of the narrative itself—namely, in the form of literal or realistic plot events. In this sense, the most interesting problem in Moorcock's novel is neither psychological nor theological per se, but rather historiographical or, in a term I promise not to reuse, psychohistoriographical. It concerns the meaning of the individual historical event and its capacity to affect and define the broader historical record, as well as, alternatively, the capacity of that historical record to define and characterize the individual event. In turn, the various psychological issues raised in *Behold the Man* are tethered to a larger, more central historiographical question, as Moorcock himself indicates superficially through his frequent allusions to Jung's theory of archetypes: what tendencies or forces motivate the historical event or are motivated by it? This question of the relationship between event and history is perhaps the central problem of modern historiography, dating back at least to Hegel. It concerns what Hayden White calls the fundamental "ambiguity of the term 'history,'" which "unites the objective with the subjective side" and "comprehends not less what has *happened* than the *narration* of what has happened."[11]

Moorcock, pursuing this historiographical inquiry within the terms of his own narrower interest in Jung, composes a series of subtle jokes about the "inevitability" of the story of Christ and about Glogauer's ironic attempts to co-opt or reinforce that inevitability:

> "There must be twelve," he said to them one day, and he smiled. "There must be a zodiac."
>
> And he picked them out by their names. "Is there a man here called Peter? Is there one called Judas?"[12]

Later, as Glogauer singles out Judas to help him—foretelling, in some ambiguous sense just as Jesus himself did in the Gospels, which one of the disciples will betray him—the ironic juxtaposition between historical fact and archetype or allegory is perfectly apparent:

> "Judas?" said Glogauer hesitantly.
>
> There was one called Judas.
>
> "Yes, master," he said. He was tall and good looking with curly red hair and intelligent eyes. Glogauer believed he was an epileptic.
>
> Glogauer looked thoughtfully at Judas Iscariot. "I will want you to help me later," he said, "when we have entered Jerusalem."
>
> "How, master?"
>
> "You must take a message to the Romans."
>
> "The Romans?" Judas looked troubled. "Why?"
>
> "It must be the Romans. It can't be the Jews. They would use stones or a stake or an axe. I'll tell you more when the time comes."[13]

A consideration of the inevitability of the past summons an ambiguity so basic to the problem of history that it cannot be contemplated without opening up, even within the course of the popular fiction, an inquiry into the ontology of the event itself. Is the "must" that accompanies the story of the crucifixion—"it must be the Romans"—constituted by (a) the sheer empirical facticity of the historical event itself (i.e., in the actual past it really was the Romans who crucified Jesus); or (b) the mythological or suprahistorical inexorability of the conflict that the crucifixion represents, a conflict between allegorical figurations of imperial-bureaucratic force and individual-istic (or monotheistic) subjectivity; or (c) the weighty exigency of the two millennia of subsequent historical events and trends, in light of which the crucifixion has been ceaselessly, but of course retroactively and even circu-larly, interpreted as an essential, archetypal cause? We could phrase the tri-lemma this way: is the historical event, in and of itself, a blankly preliminary cause, an overdetermined revisionist effect, or a mere component or signi-fier of some even larger story or allegory?[14] Or, yet again, to insert some of the language of time travel back into the three possibilities I have suggested: is the inevitability of the crucifixion the result of the somehow unalterable pastness of that event, the result of the deliberate intervention of the time traveler from out of the present (the historian? the Church?), or the result of a powerful inertia or causelike weight of history itself?

In Chapters 4 through 6, I will have occasions to discuss in greater detail the "conservative" characteristic of time travel fiction, which, perhaps sur-prisingly, tends to restore histories rather than to destroy or subvert them. This tendency toward historical conservation, like the ambiguity of the relation between present and past more generally, is a fundamental philo-sophical problem, opening up questions about the tendentiousness of events themselves and of the momentum or inertia of their histories. Most impor-tant, the time travel story efficiently conjoins questions of form, psychology, and history within the context of a fundamental historiographical query presented straightforwardly on the level of plot: How, quite literally, is the past event reconstructed by or from the present? How is it discovered, or made, to be "real"? *When* is it caused?

The time travel paradox initiated by Karl Glogauer—that the presence and significance of the story of Jesus within cultural history result from a temporal intervention out of the present—is a tale about how psychological motivations or, more concretely, masochistic and obsessive fixations, cathect, repeat, and continually reestablish the larger historical record, in essence preserving it or rendering it history in the first place. Moorcock's novel thus compels a question about the conditions within the present that drive us toward certain segments of the historical past—that is, to history's os-tensibly crucial moments. This is a radical formulation of historiography, a formulation that oversteps any lukewarm inquiry into the present's effect on the past—indeed, shortcuts the whole fraught philosophical question of "presentism"—and advances directly to the basic structure of the past it-self, its reality, as well as the connection of that reality to the ways in which stories are retold. In time travel fiction, the fundamental historiographical question—how is the past reconstructed by or within the present?—becomes a literal *topos*, is told as a tale, or is enacted by a real person seated in a vehicle or machine.

Third Reading: The Ontology of the Event in "All the Myriad Ways"

My third reading belongs to the science fiction subgenre sometimes re-ferred to as the alternate universe or alternate history story, in which the time machine transports its rider not into the past or future but rather into alternate timelines—or, to use a term from physics that also works for nar-rative theory, worldlines. In Larry Niven's 1968 short story "All the Myriad Ways," pilots flying for a company called Crosstime, Inc., take their vehi-cles to parallel universes, or "branching" worlds, and then return to the pri-mary world of the story, bringing back exotic artifacts and information about alternate histories. Part of the story focuses on the trouble that Cross-time pilots encounter when they must distinguish the innumerable alternate worlds from one another, and their invention of technological controls for accurately determining which worlds they are traveling to or from. "In those first months," Niven writes, "the vehicles had gone off practically at ran-dom," but now, since the engineers at Crosstime have improved their ma-chinery and also refined their archival and commercial goals in exploiting alternate timelines, "the pinpointing was better":

> Vehicles could select any branch they preferred. Imperial Russia, Amerindian
> America, the Catholic Empire, the dead worlds. Some of the dead worlds were
> hells of radioactive dust and intact but deadly artifacts. From these worlds
> Crosstime pilots brought strange and beautiful works of art which had to be
> stored behind leaded glass.[15]

Here, as in many other examples of the alternate universe subgenre, the language of a much older type of speculative romance—"Imperial Russia, Amerindian America, the Catholic Empire"—is directly appended to the generic language of science fiction ("dead worlds," "radioactive dust") in order to form a kind of fable about historical evolution and variation. As I discuss in Chapter 1, in older speculative fiction, especially in utopian or dystopian romance written before the turn of the twentieth century, the scientific machinery of travel itself would more likely have remained incidental or subordinate to the writer's presentation of an alternative sociopolitical milieu. However, in a science fiction alternate universe story such as Niven's, something like the opposite is true: history, culture, politics, religion, and so forth remain the background for a speculation about the specific mechanisms of movement and travel.

Niven's alternate universe story belongs to a class that arises directly from the reception by popular fiction writers of the "many-worlds" interpretation of quantum mechanics first proposed by Hugh Everett III in his 1957 Ph.D. dissertation and floated as a controversial hypothesis among physicists through the 1960s.[16] In Everett's interpretation of quantum theory, as Bryce DeWitt explains it in a 1970 gloss:

> This universe is constantly splitting into a stupendous number of branches, all resulting from the measurementlike interactions between its myriads of components. Moreover, every quantum transition taking place on every star, in every galaxy, in every remote corner of the universe is splitting our local world on earth into myriads of copies of itself.[17]

Consistent with this interpretation, the branching worlds of Niven's story are, as his title suggests, also "myriad," because the pivotal incidents that give rise to them are as indeterminately numerous as the virtually countless number of quantum events. At every moment, this theoretical model suggests— although we cannot take for granted, particularly at the quantum level, that we know even in a general sense what such a "moment" consists of—alternate histories are possible, and if possible then also real. Thus any palpable difference between two nearby branching worlds might be, from the relatively coarse perspective of a Crosstime pilot's perception, vanishingly minute: "The latest vehicles could reach worlds so like this one that it took a week of research to find the difference. In theory they could get even closer."[18] Nothing but another story, for instance another "week of research," would be sufficient to distinguish them.

From a narrative-theoretical viewpoint, it is quite interesting to contemplate two worlds that are declared, theoretically or in the abstract, to be different, but between which no concrete or practical difference can be discovered. The positing of such a pure narrative simulacrum suggests that Niven is interested (even if not fully explicitly) in theorizing the narrative event as such. He is asking, in essence, precisely what are the formal, aesthetic, and/or ontological markers of the event that could constitute a moment of narrative difference—the markers, perhaps, of fictionality, or of what analytic philosophers sometimes call "incompleteness," as a means of distinguishing between the real and its alternates.[19] Niven, in order to allegorize the ambiguity of narrative differences encountered by the Crosstime pilot who tries to return to his own worldline, continues to borrow freely from quantum theory's model of events as uncertain across a probability continuum:

> There was a phenomenon called "the broadening of the bands." . . .
> When a vehicle left its own present, a signal went on in the hangar, a signal unique to that ship. When the pilot wanted to return, he simply cruised across the appropriate band of probabilities until he found the signal. The signal marked his own unique present.[20]

In a Crosstime ship, the signal that marks the pilot's "unique present" is a clear-cut electronic manifestation, for instance a reading on a meter or gauge, or a visual or audible blip. Thus, in the science-fictional world of vehicles that use this sort of convenient technology, the pilot is able to identify straightforwardly ("simply cruise") his or her own present via the unambiguously legible mechanism of this unique marker, a positive signifier, so to speak, of the continuous and self-identical reality to which he properly belongs, his home world. However, because such signifiers in the nonfictional world are neither so present nor so unique, we are perhaps invited to be suspicious of the very idea of a signal or device that could assure one of a return to one's own reality, especially once the possibility of an infinitely divergent set of alternates has been introduced by the story itself or by the underlying physical theory upon which the story is based. Where, in the real world, could we look for signals that would distinguish an original worldline from some alternate one, or an actual one from a fictional one? Where is the signal for the originality or self-consistency of the real world?

Niven is quite cognizant of the profound problem that his fictional signal exposes concerning the artificiality and arbitrariness of narratives about reality once time travel is given free rein. Indeed, he gives the impression of being appalled by the "horrible multiplicity" that results.[21] In a later preface written for the story, Niven complains, presumably with his tongue half in cheek:

> I spent time, sweat, effort, and agony to become what I am. It irritates me to think that there are Larry Nivens working as second-rate mathematicians or adequate priests or first-rate playboys, who went bust or made their fortunes on the stock market. I even sweated over my mistakes, and I want them to count.[22]

Without presuming that Niven has read Nietzsche's *The Gay Science*, this is nonetheless a close counterpart to Nietzsche's earliest formulation of the doctrine of eternal return, presented as a thought experiment about moral decisions under the burden of an infinite proliferation of real presents: would you still choose precisely the life you are now living, even were it to be repeated throughout eternity, given your knowledge of the myriad possibilities you might have chosen instead—or, in light of such a choice, would the suddenly immense question of your decision cripple you, cause you to "gnash your teeth and curse"?[23] Like Nietzsche, Niven observes the moral crisis into which the scenario of infinitely multiplied choice plunges the conventions of narrative, which appear generally to presume that certain choices are more plausible or acceptable than others, lending themselves to stories that give at least the impression of logical or natural bias. Alternate universe stories, in their role as ontological thought experiments, effectively undercut that presumption, introducing the disturbing possibility that there is no natural criterion for preferring certain lines over others: "[E]very time you've made a decision in your life, you made it all possible ways," Niven writes; "I see anything less than that as a cheat, an attempt to make the idea easier to swallow."[24] Thus at least one ethical upshot of the alternate universe story is the death of narrative significance itself or, at the very least, the exposure of the "cheat" required to keep it alive.[25] If all narrative lines are equally possible, then the logical or naturalistic basis for realism, in the most general sense, can no longer be used to distinguish better from worse narratives, or even to pick out specifically *narrative* lines from mere amalgamations of coincidence.

I will return, at certain moments in my argument, to the question of narrative ethics, if that's what this precipitous opening of narrative possibility really amounts to, as well as to the narrative-theoretical question of fictionality. But to continue first with the more basic epistemological quandary that the apparatus of Niven's story has broached: I have been asking, what would ever correspond to a "unique signal," one that could allow one to identify one's own subjective present, one's "line," as the proper or primary world, as opposed to the branching one? What signal could enable us to call one history real and another a fiction or fantasy, even before we arrive at the apparently ethical question of many possible or plausible narrative sequences? The time travel story permits Niven to pose such questions in an especially elegant fashion. Continuing from the quotation above:

> The signal marked his own unique present.
>   Only it didn't. The pilot always returned to find a clump of signals, a broadened band. The longer he stayed away, the broader was the signal band. His own world had continued to divide after his departure, in a constant stream of decisions being made both ways.[26]

This pilot, returning from another cross-time world, encounters, in practical terms, precisely the same quandary that the narratologist or philosopher encounters in theoretical terms. Confronted with a "clump" of signals, the pilot implicitly asks what remains of the unique difference that constitutes the identity of *his* world, the one he left, or, more exactly, the one he is still ostensibly *in*, as a temporally coherent and continuous subject. At best, he is now living in a world-simulacrum, effectively or pragmatically the same as his original world only because *he does not yet know how to tell the story that would distinguish them*.[27] Yet the pilot's quandary represents that of any reader who "travels" to alternate worlds while remaining a subject within the home world of the drama of reading, and whose reliance upon the spatiotemporal coherence and consistency of that home world helps him or her to distinguish fact from fiction, reality from mere plot. There are those of us who don't make these distinctions as easily or automatically, and who may have trouble either inhabiting or discarding a supplemental story about the act of interpreting "the real," for instance the small child, the schizophrenic, the psychoanalyst, the skeptical fact-checker, or the everyday dreamer. For these more incredulous types, the seemingly natural assumption that either a subjective or objective self-identity—what Philip K. Dick calls the "inner conviction of oneness [that] is the most cherished opinion of Western Man"[28]—will survive the act of differentiating between any "real" narrative world and a hypothetical or fantastical one is not nearly so certain an assumption as we, in our narratologically coherent waking lives (or fantasies), would generally like to believe.

In essence, Niven's Crosstime pilots disclose a situation in which the arbitrary and tenuous conditions of the ostensible coherence of our "normal" world are quite literally exposed:

> Usually it didn't matter. Any signal the pilot chose represented the world he had left. And since the pilot himself had a choice, he naturally returned to them all.[29]

The worlds in proximity to the one to which the pilot returns are similar enough to his home world to render the difference between them—for there *is* always some difference, or else there would be no question of describing them as "alternates"—pragmatically indistinguishable.[30] Again one can ask, what is the criterion for this pragmatic indistinguishability? That is to say, precisely when would the difference between this and that narrated world be of *no* pragmatic consequence—like "two worlds that differ by only the disposition of two grains of sand on a beach," as Keith Laumer suggests, "or of two molecules within a grain of sand"?[31] We can only use narrative theory, either explicit or tacit, to answer this question, for the consequence of the difference constituted by a divergent narrative line is determinable only by the participant within yet another story, a participant who *notes* that difference and whose own further *narrative* of the history of difference has therefore altered his or her relation, as subject, to the events in which he or she participates.

Niven gives a single dramatic instance of such a narrated difference, one of a type that is virtually canonical within time travel literature, for reasons I will have occasion later to discuss:

> There was a pilot by the name of Gary Wilcox. He had been using his vehicle for experiments, to see how close he could get to his own timeline and still leave it. Once, last month, he had returned twice.[32]

Here is a difference sufficiently dramatic to expose a pragmatic narrative discrepancy: two Wilcoxes in the same worldline. We might therefore call the Crosstime pilot Wilcox himself a type of "laboratory narratologist"— and, indeed, Wilcox has been conducting something like field research in narrative theory, using his vehicle to try to determine the precise interval between two different worldlines, the minimum quantum of difference required to establish a pragmatic distinction. In Niven's words, he is trying to find out "how close he could get to his own timeline and still leave it." But what Wilcox seeks is basically what Tzvetan Todorov, summarizing the research of earlier structural narratologists, calls the "smallest narrative unit," or what Eugene Dorfman even more pithily calls "the narreme," the minimum divergence that could be identified as proper narrative variation.[33]

Niven, whose tale is largely premised on his own mistrust both of time travel conventions and of alternate universe stories generally, characteristically recounts Wilcox's experiment with overdramatic bravado: the Wilcox who returns twice commits suicide, as do large numbers of other people after the establishment of Crosstime, Inc. In reality (if we may continue to speak this way), the difference between narrative lines could be dramatized by events much more subtle than the return of two Wilcoxes or their suicides. In fact, it could be dramatized by any minimal event that failed to coincide with the "normal" narrative that keeps coherent, without threat of hypothetical otherness, the thread of a narrating subject. If Wilcox someday discovered, after his return, that his memory of some specific episode differed minutely from his wife's or commander's memory of it, or that some empirical fact he had always assumed to be true was now false, or that some person at the fringes of his acquaintance now failed to recognize him, we would have, in less melodramatic terms than Niven's but in terms no less philosophically significant, the same quandary, the same divergence between a subject's putatively self-identical narrative world (or "band" of indistinguishable "nearby" worlds) and a world in which a new history has been commenced and must now be renarrated. Ultimately, what Niven offers is a description of the epistemological conditions of *any* act of narrative, which commences at the moment in which the divergence between the subject's own world, on the one hand, and the hypothetical or alternate world constituted by a concrete divergence, on the other hand, must itself be renarrated.

It is therefore to his credit that Niven ends his story, not with the rather pallid psychological insight that an infinity of moral choices is equivalent to no choice at all, or is an impetus for suicide, but instead with a narratological complication that casts his plot into a kind of interworld limbo, fundamentally unfixable with respect to the specific worldline it asserts as primary, or with respect to the specific narrative closure for which it therefore opts. At one point in the story, the narrator, Detective-Lieutenant Gene Trimble, recalls that "they've found a world line in which Kennedy the First was assassinated," suggesting, of course, that the history of the *present* story is already an alternate one, in which there exists such a figure as "Kennedy the First."[34] Niven uses this brief reference to alternate political histories in order to engage in some slight humor about the 1960s American fascination with the Kennedys and their pseudoroyal "Camelot," a joke that obliquely alludes again to the prehistory of time travel literature in the sociopolitical fictions of utopian and scientific romance. In the late 1960s of Niven's story, Kennedy (the king, not the president) is still alive, and with tongue in cheek Niven declines to consider whether that alternate set of facts makes any real difference. But the brief reference to "Kennedy the First" alerts the reader to a basic uncertainty in the story concerning the precise relationship between fictional and nonfictional elements. In turn, this uncertainty anticipates a curious denouement in which a series of multiple and logically incompatible narrative lines are offered side by side:

> Casual murder, casual suicide, casual crime. Why not? If alternate universes are a reality, then cause and effect are an illusion. The law of averages is a fraud. You can do anything, and one of you will, or did.
>
> Gene Trimble looked at the clean and loaded gun on his desk. Well, why not? . . .
>
> And he ran out of the office shouting, "Bentley, listen! I've got the answer . . ."
>
> And he stood up slowly and left the office shaking his head. This was the answer, and it wasn't any good. The suicides, murders, casual crimes would continue. . . .
>
> And he suddenly laughed and stood up. Ridiculous! Nobody dies for a philosophical point! . . .
>
> And he reached for the intercom and told the man who answered to bring him a sandwich and some coffee. . . . [35]

Narrative rules, even clichés, govern whatever aesthetic or ethical choices remain to be made between the various lines juxtaposed here, even if, strictly speaking, cause-and-effect does not. For instance, in at least one of the lines, the loaded gun that appeared earlier in the story must be fired; in

another, the crime gets solved. So in the story's end—if that's what it is—
the criterion that determines the choice of closure is purely a narrative rule
or a generic convention, even as the epistemological relationship between
narrative lines and worldlines is being radically reinterpreted:

> And picked the gun off the newspapers, looked at it for a long moment, then
> dropped it in the drawer. His hands began to shake. On a world line very close
> to this one. . . .
>> And he picked the gun off the newspapers, put it to his head and
>> fired. The hammer fell on an empty chamber.
>> fired. The gun jerked up and blasted a hole in the ceiling.
>> fired. The bullet tore a furrow in his scalp.
>>> took off the top of his head.[36]

With neither epistemological nor ontological grounds for preferring one set
of events over the others—because all are equally real within the story's
myriad alternatives—we can, finally, rely only on narrative means to select
at least one of them or to place them all in a plausible order. The ontological equivalence of several contingent possibilities here comes up against
the necessity of positing a sufficient aesthetic reason for preferring one
possibility over the others. This is an old philosophical problem. As Leibniz
beautifully expresses it, the seeming contingency of such myriad possible
events, their apparent equivalence or homogeny, is only an artifact of an observer's continued "confus[ion]" about the full set of causes leading to the
single one of them that will, in the end, have remained plausible—which is to
say, in a truly "sufficient" world, possible.[37] In any fully determined (and fully
transparent) universe, were it only "distinct" to us, one worldline alone would
retain sufficient reason to exist, and therefore only one ending could result
from the infinitely interwoven arrangements of events and causes that
preceded it in the underlying story.

Of course, the question remains open whether, or to what degree, a fictional narrative is a fully determined universe, and whether, either in fiction or reality, any distinct observation of deterministic lines is possible or
even imaginable. Such questions are especially pressing within alternate
universe stories, which literalize their means (here, a Crosstime ship and a
"unique signal" on a gauge) for creating and observing multiplicity and contingency, just as a time travel paradox story like Silverberg's or Moorcock's
literalizes its means for creating narrative lacunae or historical precedence.

In this sense, Niven's story is very much like a quantum theorist's thought experiment in which hypothetical conditions of observation (or "measurement") are constructed and played out—a "Schrödinger's cat" of narration, literally depicting both the radical epistemological ambiguity of narrative alternates and the aesthetic means of their final collapse into genre.

Let us be quite precise about what is at stake in the ontological game Niven plays here. In one of the responses to Hugh Everett's "relative state" or "many-worlds" thesis, which I mentioned above, John Wheeler notes that Everett's formulation of quantum theory has the effect of eliminating any possibility of a useful "external observer" for whom "probabilities are assigned to the possible outcomes of a measurement," and who could therefore determine which result, or "line," is the most likely consequence of an observed event.[38] The status of observation in quantum theory is of course a matter of extensive analysis, and nothing to be taken for granted. However, as Wheeler suggests, the traditional Copenhagen interpretation does still posit an "ultimate" determination of that outcome "by way of observations of a classical character made from outside the quantum system"[39]—in brief, the system "collapses" onto a definite state when measured. Everett's model differs: "Every attempt," as Wheeler asserts, "to ascribe probabilities to observables is as out of place in the relative state formalism as it would be in any kind of quantum physics to ascribe coordinate and momentum to a particle at the same time."[40] Indeed, Everett's theory has the effect of further exposing the arbitrariness of the imputation of "some super-observer" who is seemingly required, within the more canonical Copenhagen interpretation, to correlate, with one of the myriad possibilities prompted by a quantum event, a *single* measurement of that event in the physical world, taken with "classical" equipment.[41] In a nutshell, Everett's "relative state formalism" or, more broadly, the "many-worlds" interpretation overall, contrives to describe the full range of myriad possible outcomes of an event, but declines ever to offer an ontological basis for adjudicating or "collapsing" them[42]—declines, in other words, to offer what Niven offers (at first) his fictional Crosstime pilot, that "unique signal" for separating and distinguishing the various potentially observed lines, and for selecting which of them to narrate as actual. In a sense, Niven's "signal" literalizes what the Copenhagen interpretation demands but Everett's formulation prohibits: a pragmatic means of super-observation, and therefore of finally *deciding* how the story is supposed to go or was supposed to have gone.

Ultimately, therefore, the interesting thing about alternate universe fiction is that, unlike the actual spacetime continuum (or perhaps more like it than we realize), it *does* offer "unique signals" for adjudicating lines of possibility, in the form of conventions of narrative structure. Among these is the vague conglomeration of half-conscious generic rules out of which both writer and reader form the sense of an ending, rules that conventionally call for guns to be fired or crimes to be solved. Although the three final lines of Niven's story are ontologically simultaneous, their specific concrete sequence in the text, and therefore their diachronic order in the plot, is nonetheless fairly strictly constrained by genre. One of the lines, namely the one that appears last on the physical page of the story, is its *de facto* ending: Gene Trimble blows off the top of his head. The sequence of *noir* detective elements that leads to this generically consistent conclusion is in turn governed by broader structural rules that tend to favor melodramatic escalations of violence near the climaxes of detective fictions. The explanation of why this should be the case might require us to traverse the entire oeuvre of structuralist narratology, with a healthy portion of genre history appended. I necessarily touch on only a small, but I hope significant, portion of that literary-theoretical landscape in this book. In a very powerful sense, the story ends as an *illustration* of itself, a visual diagramming rather than a telling of its possible outcomes. In my final two chapters, I discuss at length how and why time travel and alternate universe stories become "visual" in this way, depictions rather than narrations of their own *fabulas*. That discussion will involve, among other things, taking far more seriously and literally than narrative theorists sometimes do the structural and physical attributes of the odd term "viewpoint."

For the moment, given the specific philosophical apparatus of quantum mechanics and observability that Niven's story confronts, we may pursue this question narrowly by inquiring further about the free-indirect narrative observer who *tells* the tale, and whom the text itself exposes for severe critical scrutiny. For the quantum theorist, of course, the matter of who observes, and from what position or frame an observation is made, can never be mere postulate but must enter directly into both the calculations and the theorization of the event itself.[43] In the many-worlds interpretation, an observation remains rigorously *within* the system at hand, never outside of it, which is to say it remains precisely what it is to begin with in the history and aftermath of the quantum event: a physical fact rather than a hypothetical position or act. Hence, within the rigorously physical interpretation of quantum formalism provided by Everett and his followers, there is no possibility of a free-indirect viewpoint that could compare and contrast different lines of probable event-histories. Such a viewpoint would be precisely as unphysical as any other posited means of "collapsing" the possibilities onto a single measurement.

The unphysical "super-observer" who would (hypothetically) be capable of determining which event-histories are more probable than others, and, ultimately, which history is the correct description of the event, corresponds to the godlike viewpoint depicted in Leibniz's description of a determined universe: from *its* perspective, only a degree of unclarity about the sufficient reasons for events, an unclarity from which the super-observer alone would be exempt, creates the illusion that more than one line is possible. In the metaworld of all possible worlds, into which presumably only a god has a fully clear vista, the "unique signal" of each worldline shines out with the singular exactitude of logic itself, on an infinitely discernible continuum of possibility. But, to ask a question that anticipates my later discussion, how close is such a god to the entirely mundane position we describe as the narrator's point of view—how necessary, therefore, might such a god's perspective be to the telling or viewing of any story? The fact that the metaworld of the god's-eye view may be unphysical[44]—indeed, the fact that its inhabitation is rigorously excluded by the same philosophical framework that Niven adopts in order to construct a time travel or multi-universe story, the implications of which he finds so "horrible"—does not in the least prevent him from narrating the story *from* the viewpoint of that metaworld, in which the story of all stories, as it were, may be transmitted. "All the Myriad Ways"—and not just it, but finally any story that juxtaposes the fictional to the historical, the alternate to the real—is essentially a "super-observation" of multiple worlds, and (usually) a set of decisions about when and how to collapse them.

Contexts, Methods, Directions

I wish to mention briefly some of the fields of inquiry in which the problem of time travel has especially been of interest, and at the same time some of the contexts for the cross-disciplinary synthesis I intend to fashion in the chapters to follow.

There are two popular genres in which time travel has long played a significant role, science fiction and the romance novel. In modern romances, the time travel plot is almost exclusively a transportation medium: the hero or heroine is carried to or from a particular future or historical past, or is visited by a counterpart from that other time; some (usually) heterosexual liaison ensues. Because neither technologies of time travel, nor historicity per se, nor problems of narrative, tend to be immediately at stake in this highly regularized fiction, I will have little occasion to discuss it here, despite its considerable theoretical interest in other domains of literary studies.[45] By contrast, science fiction writers—at least after a certain historical point, as I shall detail shortly in Chapter 1—tend much more often to emphasize, over and against a political or erotic agenda, the mechanisms and significance of time travel itself, as well as its psychological, narratological, and historiographical implications. This is the case even though the viability or acceptability of time machines within the genre has been a fraught topic from at least the advent of "hard" science fiction in the late 1930s and early 1940s. Science fiction authors are divided over the generic and/or aesthetic question of whether time travel counts as proper science fiction or as "mere" fantasy, and critics have perhaps too quickly followed suit, continuing to debate whether time travel plots are legitimately "hard" or realistic.[46] Robert Silverberg himself writes in 1967:

> Among some modern science-fiction writers, stories of time-travel are looked upon with faint disdain, because they are not really "scientific." The purists prefer to place such stories in the category of science-*fantasy*, reserved for fiction based in ideas impossible to realize through modern technology.[47]

For reasons that should become clear in my brief mention of current physics to follow, as well as in my longer discussion of the origins of time travel fiction in nineteenth-century utopian romance in Chapter 1, I am not very concerned with debates over the putative hardness of time travel stories, debates that seem to me both inconsequential and obsolete alongside what I perceive to be time travel fiction's potential contribution to narratological study. The essence of time travel fiction, for the purposes of narrative-theoretical work, lies in its specific methods of constructing and juxtaposing narrative registers, layers, or lines. These methods may or may not correspond to the subgeneric or sub-subgeneric classifications of time travel stories and their paradoxes that enthusiasts (I include myself) are inclined to contemplate or dispute. Leaving such debates aside for now, the literary-critical scholarship dealing directly with time travel in science fiction is sufficiently finite to permit a summary here. I am aware of only two comprehensive books on the topic, Bud Foote's *The Connecticut Yankee in the Twentieth Century: Travel to the Past in Science Fiction* and Paul J. Nahin's *Time Machines: Time Travel in Physics, Metaphysics, and Science Fiction*, along with a number of more brief but superbly illuminating discussions by literary and film critics such as Katherine Hayles, Stanislaw Lem, Constance Penley, Brooks Landon, Vivian Sobchack, Garrett Stewart, and several others whom I will mention when their work bears on my analysis.[48]

In general, literary theorists have been relatively indifferent to time travel fiction, even where their interests unmistakably verge on the sorts of narratological questions I have attempted to raise in my three examples above. As I mentioned initially, such a dearth of attention may turn out, in future retrospect, to have been somewhat surprising. For one thing, the basic question of "fictionality," or of storytelling as "world-making," is at least as old as the theory of narrative itself, if not considerably older; similarly inveterate is the poetic conception that narratives, both fictional and nonfictional, recoup or recover past time through a kind of cerebral or conceptual "travel."[49] One might also argue that certain theoretical problems in literary modernism—for instance, Proustian *rechercher* as an immanent theorization of the structure and psychology of storytelling,[50] or Joycean and Woolfian experimentation with the spatial and temporal rearrangement of narrative forms—might have encouraged academic literary theorists to notice generic time travel fiction as an opportunity or a challenge. Of course, such notice would also require a broader shift in the object of academic literary study, from canonical texts to wholly popular and even pulp texts. This type of adjustment is proceeding apace in the academy, particularly in light of the multiplex and ironic temporalities of postmodern fiction, but is still relatively nascent, and more so in literary than in film studies. Nonetheless, it is possible to catalog, at least loosely, a series of contemporary narrative-theoretical problems that touch directly on the question of time travel, even where that question is not yet fully explicit—for instance, the problem of fictionality (Thomas Pavel, Wolfgang Iser), the problem of possible worlds and counterfactuals (Ruth Ronen, Lubomír

Doležel), the problem of worldmaking and metafiction (Kendall Walton, Mieke Bal), the problem of modality and virtual reality (Marie-Laure Ryan, Monika Fludernik), and the problem of the relation between textuality and visuality (W. J. T. Mitchell, Garrett Stewart). All the theorists mentioned in these parentheses are allied with longstanding traditions of literary narratology, traditions that are in turn traceable to the formalist and structuralist underpinnings of nearly all contemporary literary theory; a continual benchmark is the work of Gérard Genette, itself grounded in both structuralism and Russian formalism. However, it may be telling that when contemporary narrative theorists refer to specific influences on their theories, such influences are often likely to come not from either literary theory or structural anthropology and linguistics but from analytic philosophy and logic.

In analytic philosophy, time travel has enjoyed a cachet it lacks among literary theorists. Philosophers interested in problems of time, causality, and philosophical realism have very often invoked time travel scenarios as cogent thought experiments. I will have occasion to examine the technics of a few of these experiments, particularly where time travel fiction itself raises philosophical or quasi-philosophical questions of causality or paradox. For now, I will note that analytic philosophers have made use of time travel to elucidate a number of canonical problems in metaphysics and logic: causality and temporal direction (Michael Dummett, Donald Davidson), personal identity or continuity through time (Daniel Dennett, Arthur Danto), causality and realism (W. V. O. Quine, Hilary Putnam), and counterfactuals and possible worlds (Nelson Goodman, David Lewis).

An amenable touchstone for analytic philosophy is, of course, theoretical physics. Particularly with respect to questions of philosophical realism, a number of the philosophers just mentioned are directly concerned with theories of time and causality in quantum mechanics, relativity theory, and thermodynamics. Recently, however, philosophical speculation within physics itself has tended to outstrip strictly disciplinary philosophy, as a result of a burgeoning of speculations about time and causality following the work of Stephen Hawking in the 1980s and continuing with contemporary research into multiverse cosmologies and, among other topics, quantum computing. Thus, within physics itself, the once-benighted "fantasy" of time travel has experienced a surprising renaissance, and is now widely considered to be both a valuable logical exercise and a potential physical experiment. Such legitimacy is in turn reflected in science fiction, to the degree it tends to follow the lead of scientists. Where theoretical physics touches on questions of time travel and narrative, I will reference canonical work by Einstein, Bohr, Gödel, Heisenberg, Everett, and Hawking. And I will also have occasion to refer to certain more current lay depictions of physical theory, written mainly by physicists, that consider time travel both as an immediately pressing issue in itself and as an access to fundamental physical problems, for instance the interpretation of causality and the direction of time (Igor Novikov, Kip Thorne) and the feasibility of parallel universe or multiverse models (Richard Gott, David Deutsch, Brian Greene).

Genre History

I wish briefly to anticipate the unconventional, even slightly eccentric history of time travel fiction that I will be assembling alongside my analyses of specific theoretical problems within time travel stories. The development of time travel fiction has an internal trajectory that proceeds somewhat independently from science fiction overall, in keeping with its relatively iconoclastic relationship with that larger genre. As I have begun to do in this Introduction, throughout the book I will be construing time travel fiction as a certain variety of self-conscious narratological self-depiction—what I will sometimes describe as a literalization of structuring conditions of storytelling, and eventually as a diagramming or even a "filming" of such conditions. In this specific sense, time travel fiction posits or projects its own culmination—which is not the same as saying that it *ends*—at the formal extreme to which such a narratological self-depiction might be pushed. Indeed, in my second "Historical Interval," between Chapters 2 and 3, I suggest that it would already be possible to view the "closed loop" or "time loop" story, a form of paradox fiction that peaks in the early 1940s, as this culmination. I am far less concerned with the strict correctness of this revisionist historical claim, or of the watershed moment it posits, than I am with the assistance it might provide in delineating key theoretical problems within distinct periods or phases of time travel fiction. The configuration of these phases also has consequences for the way I organize my chapters in the book, so I want to sketch it briefly in advance. More detailed discussions of its implications will follow, especially in the first few chapters and "Intervals."

In each of its three phases, time travel fiction is influenced and even defined by specific developments in the popularization of science. The first phase I identify is that of "evolutionary" utopian travel, or of the "macrologue," a term I explicate in the first chapter. It runs from the late 1880s to approximately 1905. This first phase commences with the rapid burgeoning of utopian romance following Edward Bellamy's *Looking Backward*, and concludes with the almost equally abrupt decline of utopianism around the turn of the twentieth century. During this phase, time travel is always a subsidiary narrative device, utilized in reaction to certain aesthetic and conceptual demands placed upon utopian fiction by the widespread popular reception of Darwinist models of social and political development. As I argue in my second "Historical Interval," this subsidiary status is apropos even of the most famous time travel story of the period, H. G. Wells's *The Time Machine*. The prominence of Darwinism, and of biology generally, impels a general shift toward specifically temporal models of sociopolitical extrapolation: plausible utopian futures must be directly "evolved" from actual present-day conditions, not merely envisaged or conjectured as potential replacements. Under such pressures, time travel framing narratives become a valuable, possibly indispensible means for writers to link present and future realistically, and thereby to legitimize social prognostications under the rubric of evolution. In my first chapter, I examine how and why time travel becomes the default frame for evolutionary utopian narratives, and I analyze the peculiar and fruitful narrative pitfalls created in the process.

Only when the first phase of time travel as utopian macrologue is coming to an end, just after the turn of the twentieth century, does the time travel story per se begin to emerge as an autonomous type. Even so, for a long while time travel remains more fallout than innovation, an orphaned remainder of utopianism, stripped of its rationale as a bolster for evolutionary realism in romance fictions. Thus, even with its new independence, time travel fiction persists as a minor, somewhat frivolous adventure story type, often a mere comedic offshoot of scientific romance. In my first "Historical Interval," between Chapters 1 and 2, I suggest an obscure, derivative, and entirely minor work as the exemplar of this somewhat inauspicious generic origin: Harold Steele Mackaye's 1904 novel *The Panchronicon*.[51] And I propose that we consider the early stages of the autonomous time travel story as an "interregnum," awaiting a new governing cultural or scientific paradigm to replace the one lost with the waning of evolutionary utopia.

Such a paradigm arrives with the popularization of "the Einstein theory" of relativity in the 1920s, and hence the beginning of a second phase in the history of time travel fiction. What relativity physics provides, mainly, is a repertoire of new plot possibilities: temporal dilation or reversal, physical access to one's own past or future (or alternate presents), viewpoints encompassing many or all possible worlds, "narcissistic" or "oedipal" meetings, and so on. With such narrative innovations, time travel stories start to focus intensively on the multiplication or recombination of narrative lines and worlds, a focus that was nascent in earlier time travel fiction but hardly ever indulged. Time travel now becomes, above all, a literature about the forms and mechanisms of storytelling itself, or what I have called a narratological laboratory. Noticeably, also, much of the sociopolitical motivation of earlier time travel fictions is sacrificed to this intensive concentration on form and narrative structure.

When time travel writers begin fully to embrace these narratological opportunities, the already self-conscious or quasi-parodic attitude toward plots that had characterized the early genre progresses toward a certain splendid excess. The result is what is sometimes called the "time loop" or "closed causal loop" story, an invention of 1930s pulp writers in which it is impossible to determine whether a cause precedes or follows its effect; such stories become standard fodder in the pulps by the early 1940s. Overall, I identify this second period of time travel fiction as its "paradox story" phase, and discuss some primary examples of its development in the second chapter, from the early use of "the Einstein theory" in G. Peyton Wertenbaker's mid-1920s fiction to its culmination in the loop stories of the 1930s and 1940s.

The third phase of time travel, which is comparatively amorphous, I designate with the deliberately broad term "multiverse/filmic." It encompasses a range of story subtypes that follow upon the advent and triumph of paradox fiction in the midcentury. Some of these subtypes are revisionist or parodic versions of the loop story; others are quite serious forays into the psychological or narratological implications of paradox fiction; still others attempt to pursue more current physical theory, from quantum gravity to ekpyrotic cosmology. The variety of thematic and narrative concerns that occupy time travel writers roughly in the postwar period and up to the present are the subject of the final four chapters of the book, as well as the "Theoretical Interval" between Chapters 4 and 5. Let me only briefly note two of the crucial aspects of this period: first, the growth of popular time travel film and television, which has flourished more than ever in the past couple of decades, and, second, the recent scientific legitimization and popularization of multiverse physics and cosmology, which has revitalized both time travel paradox stories and alternate or multiple universe stories. The second half of my book is substantially concerned with time travel within visual media, and especially with time travel fiction's capacity to represent, or even intrinsically to theorize, problems of visual perspective and viewpoint in narrative fiction and film. Chapters 4, 5, and 6 offer readings of, respectively, an illustrated text of Samuel Delany's novella *Empire Star*, several episodes of the *Star Trek* franchise, and the film *Back to the Future*. The "Theoretical Interval" briefly discusses what I assert is the primacy of the visual in time travel narrative.

Finally, the Conclusion offers a series of suggestions about a possible "last time travel story," intended to hypothesize the trajectory of both time travel fiction itself and the narrative theorizations that follow in its wake. My final reading—a closing complement to the three readings I offer in this Introduction—is of Harlan Ellison's short story "One Life, Furnished in Early Poverty," along with the new exegetical layers generated by its adaptation for the 1980s *Twilight Zone* television series, and by Ellison's own DVD commentary. The fact that time travel fiction lends itself to such hybrid textual/visual reconfigurations is no coincidence; indeed, time travel's inherently intermediated composition is among its most provocative theoretical characteristics.

Macrological Fictions: Evolutionary Utopia
and Time Travel (1887–1905)

In the mid-1880s, when Edward Bellamy was writing *Looking Backward: 2000 to 1887*, the available repertoire of literary devices did not yet include anything so convenient as a time machine for transporting characters into the far future.[1] Lacking such a convenience, and yet plainly anxious to avoid the more fantastic types of "travel" that utopian writers sometimes used, such as supernatural visions or fortuitously discovered underground cities, Bellamy goes to considerable lengths to construct realistic and scientifically plausible means to convey his protagonist, Julian West, to the utopian Boston of the year 2000. These means include such scientific-sounding equipment as a "hermetically sealed" sleeping chamber with foundations "laid in hydraulic cement," an "outer door . . . of iron with a thick coating of asbestos," and a venting apparatus that "insured the renewal of air."[2] And although Julian's century-long sleep is eventually induced by a mesmerist, Dr. Pillsbury, a practitioner whom most 1880s readers would have immediately identified as a quack even if Bellamy did not go to the trouble of calling

him a "doctor by courtesy only," the resulting effect of the mesmerist's labor is nevertheless recounted in precise medical lingo as a "trance state" in
which "the vital functions are absolutely suspended and there is no waste of
the tissues."[3] Indeed, Bellamy's characters continually modify their descriptions of Julian's 112-year coma with quasi-scientific terms designed to play
down its miraculous qualities—a "mesmerizing *process*," "the *subject* of animal magnetism," "a *state* of suspended animation," "a *systematic* attempt at
resuscitation"—a jargon consistent with the rationalist philosophy of Julian's future rescuer, Dr. Leete, who avows that "nothing in this world
can be truly said to be more wonderful than anything else. The causes of all
phenomena are equally adequate, and the results equally matters of course."[4]
In the end, even the quackery of Dr. Pillsbury helps to render Julian's sleep
less *outré*, since naturally a mere mesmerist cannot be expected to control or
comprehend the physiological processes he might unwittingly set in motion.

Dr. Leete, who is a far more respectable physician than Dr. Pillsbury,
eventually elucidates these physiological processes for both Julian and the
reader, admixing several scientific terminologies under the rubric of a ratio-
nalistic, if rather flamboyant, cosmology:

> "No limit can be set to the possible duration of a trance when the external
> conditions protect the body from physical injury. This trance of yours is
> indeed the longest of which there is any positive record, but there is no known
> reason wherefore, had you not been discovered and had the chamber in which
> we found you continued intact, you might not have remained in a state of
> suspended animation till, at the end of indefinite ages, the gradual refrigera-
> tion of the earth had destroyed the bodily tissues and set the spirit free."[5]

It is no coincidence that the embellishments of Dr. Leete's language—"the
end of indefinite ages," "refrigeration of the earth," "set the spirit free"—
sound like the super-science of later pulp fiction, particularly time travel in
its more melodramatic modes. As I argue in this chapter, time travel fiction
first arises directly out of this kind of excessive explanatory rhetoric in utopian romance, especially where writers feel compelled to account scientifically for the astonishing narrative voyages their protagonists undertake,
and where the convolutions of their vocabularies reflect the difficulties of
providing such accounts persuasively within a realistic fiction.[6] In this
sense, time travel literature is the inheritance of what may seem, at first, to

be a merely secondary feature of utopian romance, an explanatory overflow or longwindedness, or what rhetorical theorists call "macrologia,"[7] within the frames or travel subplots that utopian writers compose. This overflow survives and persists even after the genre's more central sociopolitical motivations have begun to die away around the turn of the twentieth century. In essence, time travel fiction, in its early instantiations, is utopian travel narrative stripped of its destination, a storytelling form that has outlived its original content.

Therefore, to understand time travel, its origins and its rationale—but, more crucially, to understand the sources of its most interesting theoretical and philosophical dilemmas, and the crucial ambivalence and complexity of its relationship with other narrative types—it is essential to retrace the prehistory of the assemblage of generic fragments out of which time travel fiction eventually came to be composed after the turn of the twentieth century. In essence, I am asking two distinct but closely connected questions about late utopian romance: first, why does the problem of realistic *travel* garner such excessive or macrological attention from the writers of utopias; second, why does this same problem of travel, within utopian romance, come to be conceived predominantly in terms of *time*? The answer to both these questions, it turns out, is Darwin.[8]

Evolutionary Macrologia

Charles Darwin's *Origin of Species* was published in 1859, and *The Descent of Man* in 1871. During the intervening decade, "people became evolutionists at a remarkable speed," as Michael Ruse observes,[9] and so by the 1870s the general notion of evolution had become as exemplary of natural science as the notion of relativity was to become after the 1920s. Nonetheless, despite the meteoric celebrity of Darwin himself in both America and Europe, the enthusiasm for evolutionary theory during the latter half of the nineteenth century rarely entails any strict sense of "Darwinism," a doctrine that is usually adopted only in the form of a convenient rubric or catchword.[10] For one thing, a variety of alternative evolutionary theories continue to circulate among both scientific and popular readers, often in vigorous competition with the more orthodox Darwinian versions championed by figures such as Joseph Dalton Hooker and Thomas Huxley in England, and Asa Gray in the United States.[11] More important, even by the 1880s, when at least a nominally "Darwinistic" version of evolutionary theory has become de rigueur, the disquieting, antiteleological model of "natural selection" discernible in Darwin's actual writing still tends to get filtered through other systems of thought, such as Herbert Spencer's "synthetic philosophy," that are more compatible with longer-standing assumptions about social progressivism, anthropocentrism, and religious arguments by design.[12]

Of course, vague theories are not necessarily less likely to succeed than precise ones. By the mid-1880s, the "general idea of evolution," either regardless of or because of its distance from strictly construed "natural selection," is virtually an axiomatic groundwork for the discourse of social development spanning the political gamut from the Marxist socialism of Edward Aveling to the laissez-faire capitalism of William Graham Sumner.[13] Like "relativity" or "uncertainty" in the mid–twentieth century, "evolution," for the late nineteenth, is an elemental component of the semantics of social and cultural description, a virtually obligatory rubric even for thinkers who explicitly oppose it as a theory.

Amidst this prevailing but highly diffuse "evolutionary setting"[14] of the late 1870s and 1880s, the genre of the utopian romance achieves its greatest popular success, culminating in a flurry of publications following Bellamy's *Looking Backward*, itself one of the bestselling books in the history of American letters.[15] Evolution, as a default scaffold for theories of historical change, provides a vocabulary for explaining the social and technological advances or regressions that protagonists typically witness in the future societies they visit, and that authors often depict as the direct result of "natural selection." For instance, as Julian West tours Boston of the year 2000, his host, Dr. Leete, makes it clear that the social progress they observe results from a process that is analogous to evolution in the natural world or, more exactly, a subspecies of it. Social progress is therefore constrained by the same inexorable mechanisms presumed to govern change within all nature:[16]

> "As no such thing as the labor question is known nowadays," replied Dr. Leete, "and there is no way in which it could arise, I suppose we may claim to have solved it. . . . In fact, to speak by the book, it was not necessary for society to solve the riddle at all. It may be said to have solved itself. The solution came as

the result of a process of industrial evolution which could not have terminated otherwise. All that society had to do was to recognize and co-operate with that evolution, when its tendency had become unmistakable."[17]

Such a rhetoric of inexorable tendency and influence is entirely common within the social theorizing of Bellamy-era utopian fiction. It entails a double preconception that most utopian writers in the period seem to have shared: first, that the field of politics is seamlessly joined to a larger, cohesive evolutionary universe; and, second, that evolutionary theory itself harmonizes with the underlying paradigm of Newtonian mechanics, and operates with a corresponding degree of empirical inevitability.[18] This double harmony in turn bestows upon the entire continuum of "evolutionary" processes, from biology through psychology, sociology, and economics, the same degree of transparency and immutability that Newton's laws furnish to the interactions of physical objects: "'Man,'" as the utopian author Albert Chavannes writes in 1895, through the voice of his host/narrator, "'like everything else in the universe, moves in the direction of least resistance.'"[19] Like simple masses in motion, social tendencies just continue unless diverted by some external influence, and their rational explication therefore entails only a simple diagnostic or quantitative analysis of the relevant energies and "resistance[s]" involved. In *The Crystal Button*, Chauncey Thomas offers this account of why beggars have become obsolete in the forty-ninth century: "[Beggary] was merely a result of certain unhealthy conditions, including waste, extravagance, avariciousness, crime, and disease, which flourished in your time, and fruited and dropped their natural seed."[20] When Thomas's time-traveling protagonist asks his future host whether crime has been "abolished . . . by legal enactments," the reply is similar: "No; but we have so reduced, where we have not entirely removed, the chief inducements to crime, including poverty, excess of wealth, injustice, and ambition for undeserved power, inevitably leading to tyranny, that it is now infrequent."[21] Just as in *Looking Backward*, society is continuous with biology, so that when Thomas's nineteenth-century time traveler naïvely conjectures that social history might have proceeded by qualitative "aboli[tions]" and "enactments"—a capricious, top-down imposition of political power—the forty-ninth-century rejoinders instead offer only incremental and developmental terms: a "natural seed" that has "flourished" and "fruited." In essence, "future" social theory no longer recognizes political conflict at all ("war [and] rebellion . . . are conditions quite impossible under the present regime"),[22] but rather only a unitary, incrementally evolving, and therefore scientifically rational, chain of "conditions" and "energies" that explicitly unfolds according to the "gospel" of "the 'survival of the fittest.'"[23]

Given the fuzziness of their application in utopian romance, these same evolutionary principles, or, more precisely, these metaphors largely standing in for principles, can be invoked to support diametrically opposed social ideologies, at times with ironic results.[24] For instance, in William Dean Howells's *A Traveler from Altruria* (1894),[25] the Altrurian guest to nineteenth-century Boston, a city that exemplifies, to this visitor, a "waking dream" of competitive conditions "that we outlived so long ago," is shocked to discover that social ranks are still determined by "a process of natural selection."[26] Nevertheless, despite the Altrurian's disdain for Darwinism, he proceeds to describe the turning point in Altrurian political history as "The Evolution," a sea change by which men and women were finally "freed from the necessity of preying upon one another."[27] Here, Howells invokes Darwinism as the very mechanism by which a Darwinistic society is eventually abolished, as though no discourse except evolution were available to underlie even directly competing social theories.[28] And this is Howells, a very subtle thinker and writer;[29] in the hands of less ingenious authors, the pressure of evolution is great enough to gainsay even straightforward logic, as when W. W. Satterlee invokes Darwinism so reflexively that it can authorize, virtually simultaneously, both free market rapacity and quasi-Christian altruism:

> Among these thronging masses I saw here and there men and women who kept sturdily on their way. The opposition which they met was their strength, the antagonism brought the possibility of victory. All they asked was the freedom to do their best, the only absolute law they knew was the survival of the fittest. These were noble souls and as they passed upward they stooped to the right and to the left to help a struggling brother.[30]

Thus "survival of the fittest" must apply despite all incongruity: "noble souls . . . help a struggling brother" at the very moment they trample him in fidelity to the "absolute law" of competition.[31]

Reading such conceptual hodgepodges, one may surmise that during the closing decades of the nineteenth century, perhaps for nearly the last time, a variety of explanatory models based on the interplay of forces—Darwinism, Marxist (or other) socialisms, industrial management, macroeconomics, the various burgeoning fields of professional engineering—could still appear potentially to belong to a single unified science, simultaneously natural, psychological, and sociological, yet wholly in accord with the postulates of evolution, which would itself be graspable as a direct extension of Newtonian mechanics. "[I]n society," the well-known author Benjamin Kidd confidently states in 1894, "we are merely regarding the highest phenomena in the history of life, and . . . consequently all the departments of knowledge which deal with social phenomena have their true foundation in the biological sciences."[32] As such, Kidd writes, they are governed by "inexorable natural law."[33]

The Macrologue and The Crystal Button

The examples cited above show how profound and productive, if not always fully constructive, the effects of evolutionary discourse are upon utopian writers during the Bellamy era. "Darwinistic" thinking compels unprecedented realism in the prognostications of utopian romances, which glean from evolutionary theory both protocols and prerequisites for depicting sociopolitical possibilities. Indeed, because late-nineteenth-century utopianists are dealing directly with models of social development, the evolutionary thinking that undergirds the incipient fields of sociology and political science is embraced with greater urgency in this literature than perhaps in any other. To ignore Darwinism, for a utopian writer toward the end of the nineteenth century, is to risk creating the impression of ineptitude or obsolescence in one's narration, and therefore to hazard discrediting one's own political theory in advance, a hazard that must have attached to more general anxieties about appearing scientifically naïve. Nonetheless, once this evolutionary will-to-realism begins also to inflect the framing plots of utopian romances, it generates delicate, possibly insurmountable challenges to realistic *narrative* structuring. If a strictly evolutionary utopia is to be visited within a compelling fictional world, as opposed to a mere treatise or broadside, then only plausibly *scientific* means of travel to such a utopia will suffice.

But this also means that by the time of *Looking Backward*, almost as a stipulation, up-to-date utopias must be set in the future.[34] Any creditable utopian (or dystopian) society informed by Darwinistic sociopolitics must extrapolate its conjectured *polis* from actual present social conditions, since that *polis* will necessarily have evolved precisely from them. A utopia can no longer be situated in a fantastical setting or alternative locale such as a lost island, distant planet, or subterranean crevasse, all milieux that by the late 1880s have become significant hindrances to plausibility.[35] Indeed, utopias very likely can no longer be "spatial" at all.[36] Instead, scientifically realistic utopian romance, consistent with the paradigm of evolution and its continuity with physics and mechanics, is set in *our* future. In turn, the conventional utopian protagonist's travel, within the frame of the narrative, must be strictly temporal, not spatial, because no matter how historically distant or evolved a utopian society is, nevertheless it must still be located "here."

We now confront the fundamental irony of late-nineteenth-century romance narration, and the dilemma that both destroys and preserves its chief aesthetic features. The pressure to provide an evolutionary social future, one that is fundamentally realistic according to the parameters of current political theory, also vastly increases the difficulty of rendering the narrative itself realistic, since by necessity time, not space, must be traversed in the story. By the mid-1880s, the requisite temporal travel is generally achieved very directly, if still problematically, via long sleep or suspended animation. Even as early as the mid-1870s, utopianists who have not yet explicitly adopted temporal frameworks for their tales are beginning to make clear that geographically distant or "lost" utopias nevertheless *represent* versions of actual social evolution, arising from conditions directly parallel to those of contemporary America or Europe, and that even these quasi-distant utopias therefore ought to be viewed as equivalently derivable from our real present.[37] This general tendency toward temporally conceived realism brings us back to the problem with which I began the chapter: Bellamy's macrological efforts to ground and legitimize the extraordinary means required to transport his character to a real future Boston.

Indeed—and this is really the key insight to be gleaned from the entire constellation of historical and aesthetic transitions that eventually connect the utopian romance to time travel fiction—the macrological tendency of utopian fiction, a longwindedness noted by nearly all of its critics, and that becomes especially palpable in its travel subplots, now virtually takes on a life of its own, becoming identifiable as the emergent precursor of what we would recognize today as the time travel story. In essence, the macrologia of utopian travel solidifies into a distinct literary form or frame, the nascent shape of a subgenre. Jean Pfaelzer usefully describes the ambiguous structure of nineteenth-century utopian fiction, with its uneasy melding of romance and sociopolitical preaching, as "apologue," a term well designed to capture the inevitable but uneasy amalgamation of the two basic elements of this genre, "fable" and "manifesto."[38] In light of my discussion of the evolutionary pressure on utopian realism, I propose we identify a new distinct component within utopian structure, the "macrologue," a counterpart to Pfaelzer's bipartite "apologue." The macrologue is that portion of utopian fiction which contains any and all efforts toward framing the requisite travel to a realistic utopian future, and therefore all of the (macrological) explanation required to *realize* that travel adequately. It is the macrologue alone that will survive utopian romance's demise and that will become, eventually, the time travel story.

There is no better illustration of the new autonomy and salience of the utopian macrologue than the strange publication history of Chauncey Thomas's *The Crystal Button*, a book I have already cited for its paradigmatic use of Darwinism and other scientific vocabularies. Thomas had apparently written a complete version of *The Crystal Button* in the early 1870s and had presented it to the publisher George Houghton in 1880. At that point, the work consisted of a conventional utopian apologue, resembling the allegorical stories typical of utopian fiction through about the mid-1870s, most obviously Samuel Butler's *Erewhon* (1872). In an editor's preface added to the later published version of *The Crystal Button*, Houghton recalls that when he first read Thomas's manuscript in 1880, he thought the work "needed and well merited somewhat more finish, and also required to be sustained by some sort of narrative."[39] This response discouraged Thomas, who laid the manuscript aside. However, by 1889, when the success of *Looking Backward* had unexpectedly revealed the marketability of utopian tales, Houghton and Thomas renewed their interest in the work and set about collaboratively revising and updating it, intending to "gather [it] together in the form of a connected narrative."[40]

What Houghton and Thomas chiefly did, seemingly aware of the inelegance of this strategy, was to attach a new framing narrative to the existing apologue, in essence an entirely new macrologue more in keeping with current literary trends: a "slight story . . . cut in two and used as 'Introduction' and 'Conclusion.'"[41] As part of this new frame, Thomas includes a

Bellamy-like "Postscript," written in the voice of the narrator (or nominal author), Paul Prognosis, explaining how the temporal anomaly or "dual life" that the new framing narrative introduces into the story might be scientifically justified.[42] This macrologue describes how Paul, while trying to rescue a drowning co-worker from a freezing river, is himself plunged into the water, where he "no doubt received a severe blow on the head from the ice or a drifting log."[43] Paul then "awakes" to find himself in the year 4872, where his host, Professor Prosper, engages him in an extended conversation typical of utopian apologues. All the while, as the newly appended macrologue informs us, Paul (or at least his body; this is never made fully clear) remains back in his bed in the 1890s, sunk in a semi-comatose, hallucinatory state from which he will emerge only ten years later.

The macrologue tells us that Paul's periodic unconscious ramblings—or so they appear to his family and doctor throughout Paul's lengthy coma—are actually snippets of the ongoing conversation he is having with Professor Prosper in the future, all about the "city of Tone," the technologically advanced successor to "ancient Boston." But Thomas goes to some effort to assure us that, if Paul is all the while dreaming or hallucinating—since Thomas and Houghton did *not* alter the manuscript sufficiently to remove that outmoded idea, any more than they revised the old-fashioned allegorical character names—the reader may nonetheless reconceive this dream in scientific-sounding terms, for instance as a "projection" of an actual future experience. Here is one of Thomas's descriptions of Paul's semi-coma within the new macrologue: "There was life still in there. And why not? If nothing material can be utterly destroyed—not even the delicate fabric of this rice-paper, which burns and leaves no ash—how much less should we expect to see the immaterial blotted out of existence."[44] This vaguely Cartesian argument, that the immaterial is more durable than the material, then gives way to a quasi-hydrological theory of the relation between dreaming and (projected) reality, a theory that appears to be grounded in mechanical engineering, and in which, even more oddly, what is immaterial "materialize[s]," but only when stripped of its materiality:

> Might not the swift current of his mental activity, accidentally diverted from
> its normal confines, have made for itself an underground course, where no eye,
> however sympathetic, could follow its secret windings? Might not his former
> projects in the realm of mechanics, and his prophecies that others had consid-

ered wild fancies,—might not these, when no longer fettered by limitations of matter and mechanical means, have finally materialized?[45]

Paul himself is an engineer, or at least a "mechanical expert," as the new macrologue pointedly reminds us, and somehow this vocation contributes to his ability to forecast the future of Boston in the guise of a "material" reality, even while "the current of his thought is broken." "Might not," Thomas's macrological narrator asks, again in his characteristic hypothetical mode, "his *could be* of yesterday have become the *now is*?"[46] Even the newly appended postscript, in which Paul describes how he "continued to live two lives," proposes a number of quasi-scientific ambiguities and gaps concerning "my mental aberration, leading a life among the scenes here described that had all the apparent reality of life in the material world," so as to lend greater verisimilitude or "materiality" to the otherwise too-fantastic motif of the utopian dream vision.[47]

The Macrologue after Utopia

Thomas's illogicalities, even (or especially) when they border on the absurd, reflect the incompatibility of the various aesthetic and philosophical demands with which the utopian macrologue is now forced to contend: a scientifically plausible form of temporal travel, a direct evolutionary link to an actual, not a merely hypothesized, future, a detailed and credible sociopolitical theorization, and a novelistic realism. All of these demands abide alongside the basic formal conundrum of a narrative that, although supposedly composed in the future, nonetheless presents itself as an extant "manuscript" in the present. Ultimately, within such examples, it becomes clear that the macrologia of utopian fiction, its predilection or obligation to overexplain, is beginning to explode into incoherence, or even into unintended parody, at the very moment it verges on a distinct and self-sufficient generic life. Bellamy's influence had been so sudden and pervasive, and utopian romances subsequently produced at such a rapid rate, that very quickly, during the 1890s, the utopian macrologue starts to transmogrify into a kind of self-standing joke, a more or less deliberate satire about the obsessively explicated plausibility of the time travel frame-narratives that must have relentlessly confronted readers during the period. As early as 1893, it becomes difficult to decide whether, for instance, the setup of a book like Frank Rosewater's *'96: A Romance of Utopia* is parodic or not. Rosewater himself seems unsure. Bolstered by technical claims about time and space being "fields," the protagonist's friend Dr. Giniwig describes his invention of a new "shampoo" that, when applied to his patients' heads, allows them to time travel, a pharmacological analog of the mesmerism or long coma more commonly used for this purpose in prior utopian romance: "I have so far succeeded that by acting upon the cellular corpuscles of the brain I transpose my patients back ten, twenty, nay fifty years in time, making the old become young and the sick return to primitive health."[48] But shampoo is only the first of numerous travel devices that Rosewater employs in his apparent obsession with framing and explaining. There is also a balloon voyage, a lightning storm, a sudden arrival in a Gulliveresque country simultaneously displaced in time and miniaturized "from some unknown cause," and a mysterious migrainelike "blotch" that appears before the protagonist's eyes and initiates his temporal adventure.[49] One must assume that this "blotch" is a metaphysical one, since apparently it is generated by the protagonist's incapacity to realize that events in time—specifically, the delivery of a cablegram informing him of a severe downturn in his finances—are fundamentally mutable "illusions" or "hallucinations" that, as Rosewater suggests, can be altered and corrected by the time traveler, given the right attitude.[50]

None of this material is even remotely well integrated or motivated; Everett Bleiler calls it "almost unintelligible."[51] Similarly conglutinated framing medleys can be discovered in many other romances of this period, for instance Will N. Harben's *The Land of the Changing Sun* (1894), in which an absurd collection of frames—an abortive balloon flight with "the professor," a vague speculation that the lost travelers have entered some land in which a new sun "revolves round the earth from north to south and dips in once a day at the north and south poles,"[52] and finally a captive voyage in a crystal-windowed submarine (along with other futuristic marvels on display)—all intervene between the book's first page and the commencement of the utopian apologue within the futuristic society of "Alpha."[53] What is perfectly clear, in the self-revisionist stage that the utopian romance has entered with writers such as Rosewater and Harben, is that the macrologue has taken on a life of its own, quite independent of its prior role as frame for the utopian apologue. More precisely, the macrologue has now asserted, beyond any possibility of repeal, its own incongruity or incoherence, thereby also becoming a chief repository of writers' narrative efforts and energies, and far more than a merely instrumental device for utopian tales. In this new augmented role, the macrologue, ever more preposterous, gets stitched together out of whatever scientific or pseudoscientific discourses its author finds at hand for travel, within the broad constraints of his or her particular revisionist skills, and increasingly with an awareness that both the realism and the science of time travel have now become, or perhaps always were, basically incoherent, and not to be taken entirely seriously.

As the utopian romance advances toward its final moments, which will arrive soon after the turn of the twentieth century, even writers considerably more adept than Rosewater or Harben have a tough time composing coherent macrologues, which increasingly exhibit the quirks of a fundamentally unsettled but now autonomous subgenre. By 1905, when H. G. Wells's *A Modern Utopia* appears, continued efforts to frame utopian romance with any semblance of realism now verge on outright parody—"frame-breaking" or "meta-utopia," in Gary Morson's terms.[54] The macrologue of *A Modern Utopia* possibly represents the limit to which the practice of travel exposition within utopian narrative—including Wells's own prior efforts, such as *When the Sleeper Wakes* (1899)—might be pressed. Not only is the entire body of the text recounted as a sort of tenuously speculative hypothesis, but the reader is first offered, in a discrete chapter typeset in italics and entitled "The Owner of the Voice," an entire theory of the nested authorial figures he or she will be required to negotiate in order to comprehend the intricate standpoint that the subsequent apologue plans to offer up:

> *Throughout these papers sounds a note, a distinctive and personal note, a note that tends at times towards stridency; and all that is not, as these words are, in Italics, is in one Voice. Now, this voice, and this is the peculiarity of the matter, is not to be taken as the Voice of the ostensible author who fathers these pages.*[55]

Furthermore, as Wells acknowledges with some humor, the "stridency" of the Voice is a necessary evil within utopian romance, and cannot be reconciled with any novelistic intentions or aesthetics:

The entertainment before you is neither the set drama of the work of fiction you are accustomed to read, nor the set lecturing of the essay you are accustomed to evade, but a hybrid of these two. . . .

. . . The image of a cinematograph entertainment is the one to grasp. There will be an effect of these two people going to and fro in front of the circle of a rather defective lantern, which sometimes jams and sometimes gets out of focus, but which does occasionally succeed in displaying on the screen a momentary moving picture of Utopian conditions. Occasionally the picture goes out altogether, the Voice argues and argues, and the footlights return, and then you find yourself listening again to the rather too plump little man at his table laboriously enunciating propositions, upon whom the curtain rises now.[56]

For *A Modern Utopia*, the macrological tendency of utopian fiction is so aggressively apparent, and so blatantly artificial, that simply to set up the conventional content or apologue of a utopian story requires a blatant extremity of framing. The macrologue here is, as it were, the filmic representation of the apologue, essentially a "cinematograph," in Wells's own term, uneasily reproducing a re-visioning of itself. Its closest resemblance is perhaps to the outlandish metafiction of Beckett's *The Unnameable*, in which the titular pseudocharacter sits in an indeterminably vast "place" or "pit," watching as other characters from Beckett's prior novels orbit around him "at doubtless regular intervals" and "always in the same direction."[57]

For Wells, by 1905, the basic sociopolitical faith that had perhaps artificially sustained the will-to-realism of the Bellamy-era utopia is now fully eviscerated, and what remains is largely the *form* of utopian romance narrative, animated by the mere residual inertia of the macrologue and its hybridized scientific rhetoric. In fact, that form is now poised to begin a new afterlife as time travel fiction, as perhaps no author was better disposed to realize than Wells himself.[58] Reviewing the prehistory of time travel fiction via the precipitous rise and fall of the evolutionary utopian romance—the primal genre of the realist long sleep—we may observe a mode of radical temporal narration congealing out of a subsidiary framing motif and into a primary story type, while the original thematic and sociopolitical motivations for its subordinate role fall away. The time travel story is the residue of utopian visitation once the destination of utopia is forsaken—the *travel itself*, minus its political *raison d'être*—or a newly unalloyed form of meta-narrative movement, a story about the structural time and space of story-telling itself.[59]

The First Time Travel Story

I have traced the lineage of the early time travel story to another genre's residual component, the macrologue of utopian romance. In offering such a genealogy, I have gone against the more usual tendency among literary critics, as well as among historians of science fiction, to locate the origins of time travel in the literary inventions of one or two specific authors, most often H. G. Wells's *The Time Machine* (1895) or possibly its earlier version, "The Chronic Argonauts" (1888). Other critics point to Mark Twain's *A Connecticut Yankee in King Arthur's Court* (1889), Enrique Gaspar's *El anacronópete* (1887), or Charles Dickens's "A Christmas Carol"—or, with an even more discerning or arcane historicism, identify originary time travel in minor works such as F. Anstey's *Tourmalin's Time Checques* (1891) or Edward Page Mitchell's anonymously published story in the *New York Sun*, "The Clock That Went Backwards" (1881). These empirical accounts of time travel genre, like any positivistic literary history, may tell us something about who wrote possible antecedents of later time travel tales, and when and where they wrote them, but little else about the more complex literary and extra-literary influences that prompted a given writer's devices, or that initiated the narratological concept of time travel itself. Moreover, such critical positivism proves particularly unsatisfactory with respect to the odd history of time travel fiction, which inherited its most interesting aesthetic and theoretical tribulations from the fits and starts of other types of fictional and scientific thinking, and whose lack of a single originating moment is an essential facet of its lasting adaptability and influence as a narrative mode.

As a genre or subgenre, time travel fiction did not begin with someone's idea, not even with Wells's or Twain's, but rather precipitated out of the partial failures of several other literary types. It is less an invention than an accommodation to a variety of mutually incompatible aesthetic, scientific, and social pressures that, during the last two decades of the nineteenth century, both produced and destroyed its immediate literary precursors. In essence, the time travel story is what lingers on after a fortuitous malfunction or mutation, a kind of fallout from the implosion of utopian fiction under the weight of its own aesthetic and political contradictions. And it is only as such—which is to say, not as an idea but rather as a formal and structural precipitation or coagulation—that time travel is nascent narrative theory.

In light of the more dialectical, and messier, history of time travel fiction I am proposing, let me offer an alternative, in fact a rather antithetical, suggestion for the first time travel story: Harold Steele Mackaye's *The Panchronicon* (1904). This book is sufficiently obscure in literary history that no one, I presume, could mistake it for an authorial coup of genre formation; moreover, it has the advantage of being decidedly derivative, even a symptomatic or parasitic work. *The Panchronicon* might initially be described as one of the earliest post-utopian time travel adventures, fully versed in the macrological framing conventions of utopian travel, yet at the same time willing to acknowledge or even poke fun at the obsolescence of the sociopolitical ambitions that prompted and supported such conventions. Mackaye's book is therefore, perhaps by necessity, a comic parody of utopian romance or, more specifically, a parody of the utopian macrologue, to a degree that even Wells, in his most self-conscious later utopias, never achieves. Parody is the mode appropriate to a literary interregnum, a period in which time travel fiction, following its adolescence as the narrative *means* of utopian romance, still anticipates maturation into a distinct story type, or still awaits the scientific scaffold, coming soon enough with the popularization of Einstein, that will support its more fully fledged literary endeavors. *The Panchronicon* is therefore a first inkling of time travel's second phase, which will begin to solidify around the concept of relativity in another fifteen years or so.

In Mackaye, we see the familiar elements of utopia recast solely in the form of a novelistic yarn, denuded of their political force or faith. The book begins with the familiar macrologue designed to explain the means of travel, in this case presented as a chapter entitled "The Theory of Copernicus Droop." Droop himself is a dissipated, alcoholic caricature of the Yankee engineer of some earlier romances, a kind of decadent pastiche of Twain's Hank Morgan, Bellamy's Julian West, and Wells's anonymous "Time Traveller"[1]—along with a healthy dose of Dr. Leete, since there will be no supplementary host-explicator where Droop eventually journeys. His fellow travelers, to whom he exposits his theory at length, are Rebecca Wise, a sensible spinster, and her younger sister, Phœbe, "the beauty of the village."[2] Already in the comically allegorical names of these characters we detect a rediscovered nonchalance about the strict realist protocols by which Bellamistic utopias are generally constrained. Mackaye clearly intends this matchup of the Wise sisters with the eccentric Droop to be a slapstick revision of allegorically inflected fixtures originating somewhere between an older Butleresque utopian fiction and the contemporary realist novel.[3]

Droop invites the Wise sisters to invest money in his "panchronicon," a vehicle that will allow him to go backward and forward through time at an accelerated rate. In a move that anticipates many later pulp time travel stories, Droop's chief motive is described as neither scientific investigation nor social reform but rather petty profit, as he proposes to return to the "centennial year" of 1876 in order to sell "future" inventions there. Nonetheless, despite Mackaye's pointedly ridiculous descriptions of science (and its potential for profit), we can detect the residues of a hybrid and holistic evolutionary discourse at work. The mechanism of time travel Droop outlines is a clear parody of discourses pervasive within time travel macrologues until only a few years earlier, and plays upon both the plausibility of the language of physical sciences and the absurdity of the conclusions that may be sanctioned by that very appearance of plausibility. This combination is bolstered by the assumption that Droop's eccentric musings will be naïvely accepted by his unsophisticated female partners, while the (presumably male) reader will better comprehend both the science and the parody:

> "Now don't you go to think I'm tight or gone crazy. You'll understand it, fer you've ben to high school. Now see! What is it makes the days go by—ain't it the daily revolution of the sun?"
>
> Phœbe put on what her sister always called "that schoolmarm look" and replied:
>
> "Why, it's the turning round of the earth on its axis once in—"
>
> "Yes—yes—It's all one—all one," Droop broke in, eagerly. "To put it another way, it comes from the sun cuttin' meridians, don't it?"
>
> Rebecca, who found this technical and figurative expression beyond her, paused in her knitting and looked anxiously at Phœbe, to see how she would take it. After a moment of thought, the young woman admitted her visitor's premises.[4]

Droop then suggests that a ship that could "cut meridians" at a faster rate than the Earth itself would be able to accelerate either forward or backward through "the days." The internationally standardized time zones were still a recent invention in this period, and the notion of an "international date line" which, when crossed, seemingly gained or lost the traveler a day, was itself a new idea.[5] In this sense, the book takes precisely the same advantage of up-to-date "science" as utopian or scientific romance, applying it to what is clearly already perceived as a fantastic or super-scientific idea, time travel. In this duplicitous use of scientific explanation, we perceive hints of the debate that will later preoccupy science fiction writers and critics over the "hard" versus the fantastic nature of time travel narrative. Mackaye continues to interlace discourses of scientific fact, particularly astronomy and geography, with a number of terminologies that, by 1904, could be perceived as generic and hackneyed, such as a Twainesque language of the Yankee inventor's blithe ingenuity, and a hyperbolic vocabulary of evolutionary time scales borrowed from more politically committed utopian writers such as Bellamy, Wells, or Howells. For instance, Droop proposes to speed the time travel process by taking his ship to the North Pole, where he will be able to "cut meridians" at a more rapid rate:

"But what if ye go to the North Pole? Ain't all the twenty-four meridians
jammed up close together round that part of the globe? . . .
. . . "Then ain't it clear that ef a feller'll jest take a grip on the North Pole
an' go whirlin' round it, he'll be cuttin' meridians as fast as a hay-chopper? . . .
[A]in't it sure to unwind all the time thet it's ben a-rollin' up?"[6]

This "theory" is tested and proved correct by the three protagonists, who
accidentally travel back much further than 1876, in fact to early modern
England, where, among other things, they persuade Francis Bacon to take
credit for writing Shakespeare's plays. Here, with the protagonists journeying into a mythologized and farcical medieval past, very little of the sociopolitical interests of the utopian or dystopian romance can still be discerned,
nor of the didactic or "Socratic" mode of conversation so much noted or
lamented by critics of the utopian apologue. What Mackaye offers instead
is essentially an action-adventure novel, in a mode that will increasingly
become the province of adolescent fiction.

With such work, the interregnum in the history of time travel begins. At
its other end, time travel fiction will finally emerge as a popular literary
type, fully matured out of its merely auxiliary and parodic modes, only in
the era of Hugo Gernsback's pulp science fiction magazines, following the
end of the Great War. The catalyst for this shift into time travel's second
phase—and for a species of time travel story that takes itself seriously as "scientific fiction" and is no longer either a macrological subsidiary or a caricature of another genre—is the emergence of the "Einstein theory" of relativity,
and of Albert Einstein himself as an international celebrity, following the
famous solar eclipse of the year 1919, during which certain aspects of the
theory of relativity are very publicly confirmed. It is to these events I turn in
Chapter 2.

TWO

Relativity, Psychology, Paradox:
Wertenbaker to Heinlein (1923–1941)

In 1907, in a paper anticipating some aspects of his forthcoming general theory of relativity, Albert Einstein suggests that light rays approaching a massive body such as the Earth "are bent by the gravitational field" of that body. He then adds, regretfully, that "the effect of the terrestrial gravitational field is so small . . . that there is no prospect of a comparison of the results of the theory with experience."[1] But four years later Einstein is more sanguine:

> I have now come to realize that one of the most important consequences of [my former treatment] is accessible to experimental test. In particular, it turns out that, according to the theory I am going to set forth, rays of light passing near the sun experience a deflection by its gravitational field, so that a fixed star appearing near the sun displays an apparent increase of its angular distance from the latter, which amounts to almost one second of arc.[2]

Of course, any "experimental test" of the sort that Einstein proposes would depend upon an observer's being able actually to see and distinguish those fixed stars that appear closest to the sun. While impossible on a normal day, such viewing may be done under special circumstances:

> Since the fixed stars in the portions of the sky that are adjacent to the sun become visible during total solar eclipses, it is possible to compare this consequence of the theory with experience. . . . It is greatly to be desired that astronomers take up the question broached here.[3]

Astronomers do take up Einstein's question in 1917, when Britain's Astronomer Royal, Sir Frank Watson Dyson, announces that an especially convenient total solar eclipse will occur a mere two years hence, on the date of May 29. It is "the most favourable day of the year for weighing light," as the physicist (and secretary of the Royal Astronomical Society) Arthur Eddington remarks, because "the sun in its annual journey round the ecliptic . . . is in the midst of a quite exceptional patch of bright stars— part of the Hyades—by far the best star-field encountered."[4] To prepare for this auspicious confluence of circumstances, a "Joint Permanent Eclipse Committee" of the Royal Society and the Royal Astronomical Society is formed to plan expeditions to view the eclipse. In 1917, the Great War is still raging, and any plans for overseas scientific research are provisional at best. However, by early 1919 the fighting is effectively over, and travel and shipping are more easily contemplated. That spring, Frank Watson Dyson embarks for Sobral, Brazil, and Arthur Eddington for the Isle of Principe off the coast of West Africa, each with a team of astronomical researchers prepared to observe the May 29 eclipse and test Einstein's prediction.

The expeditions fulfill their promise, and five months later, back in London, Dyson and Eddington present their findings to an assembly of the Joint Committee. Photographs and measurements taken during the eclipse appear to demonstrate that the sun's gravity has indeed "bent" the light rays of the Hyades stars; or, in the terminology that these experiments are intended in part to validate, those light rays have pursued a "straight" (geodesic) path through a region of space "curved" by the sun's gravitational field.[5] In either set of terms, the path of starlight is shown to have been deflected by "the amount demanded by Einstein's generalised theory of relativity."[6] Thus, sensationally and lucidly, the "Einstein theory" receives, in Hendrik Lorentz's words, its "striking confirmation."[7]

The impact of this news is sensational and widespread, and it promptly establishes the international renown of both Albert Einstein himself and theoretical physics generally. In a number of ways, the popular cultures of both Europe and the United States are well disposed to embrace a figure like Einstein. The headlines about the eclipse declare scientific revolution— "Shifting Foundations of the Universe"; "Great Achievement"; "The Most Sensational Event in Physics Since Newton."[8] But inevitably, the reportage turns to iconic personal description: "a man in a faded gray raincoat, topped off by a flopping black felt hat, which nearly concealed straggling gray hair that fell over his ears," as the *Chicago Tribune* describes Einstein on his first appearance in the United States;[9] "a very simple, a very modest, and very great man, one of the greatest thinkers of the last 500 years," *The Manchester Guardian* states during Einstein's visit to London.[10]

Technological advances of the Great War—the development of machine guns, U-boats, and chlorine gas—are often credited with disabusing both Europeans and Americans of a variety of naïve conceptions of scientific progress.[11] But sacrificed along with these naïvetés, perhaps somewhat less obviously, is the fantasy of the independent or amateur scientist/inventor, the Edison, Bell, or Wright brother, a figure superseded by the more mundane professional scientist, whose research is inextricably tied to the institutions and vocabularies of business, government, and social planning. Within such a context, popular depictions of Einstein emerge as especially iconic and resonant, a nostalgic amalgam of pure thinker and eccentric tinkerer, an embodiment of the "inventor mythos."[12] Thus the quasi-mythical character of the brilliant reclusive scientist—for instance, the humble patent clerk who, having failed his university entrance exams, might nevertheless revolutionize the concepts of space and time in his spare hours—achieves its apotheosis at perhaps the same moment at which pragmatic options for performing and publishing scientific research are pushed inexorably beyond the purview of amateurs and into the university laboratory, the military budget office, and (eventually) the corporate research and development division.[13]

The circumstances under which Einstein's success erupts into public view might give us clues as to how physics itself will soon percolate into popular literary and visual culture, becoming the basis for a second phase of time travel fiction in the 1920s and 1930s. Certainly, a newly prevalent metaphorics of the mutability of space and time, and of the multidimensional spacetime continuum, serves as fodder for fictional speculation. Basic implications of relativity theory, presented by Einstein himself as well as by a series of popularizers with varying degrees of fidelity to the underlying physics or mathematics, prove very fruitful for "scientific fiction":[14] that time "goes more slowly" when velocity increases (and that it would "stop" at the speed of light); that bodies "suffer a shortening in the direction of [their] motion"; that their mass is likewise proportional to their velocity; that our universe is—not only as a hypothesis or thought experiment but objectively and even pragmatically—a four-dimensional spacetime manifold; that a body's potential energy is a function of its mass, and vice versa ($m=L/c^2$, or in its more common version, $E=mc^2$); even (dubiously) that "effect may precede cause."[15] All of these potential consequences of relativity theory had been known for some time, many dating back at least to Lorentz and Fitzgerald's interpretation of the Michelson–Morley experiment in the late 1880s, or to Einstein's own publication of the special theory (1905) and general theory (1916). However, not surprisingly, it is less the theory itself than the sum of its quasi-visual poetic descriptions— amazing yet seemingly comprehensible vignettes of bending light, slowing clocks, and four-dimensional travel—that inflects the popular and literary interest in physics from the 1920s through the 1940s.

My concern, for the immediate purposes of a brief cultural history of time travel fiction, is to ascertain how specific innovations in physical theory become material for writers and readers who construct fictional worlds and narratives. Time travel, at this pivotal moment early in the 1920s, is poised to begin its second ascendancy on the heels of the Einstein theory, having lain somewhat dormant during the period I call its "interregnum." Here, a notion I suggest in the first chapter about evolutionary biology may also be true of relativity physics: a vague theory is at least as likely to succeed popularly as a precise one, and strict physical and mathematical rigor are rarely, in themselves, criteria for poetic or aesthetic utility. Hence the impact of relativity upon time travel stories is perhaps different from what one might expect. Within time travel fiction, the inheritance of the Einstein theory, for all its vast importance within the physical sciences themselves, is not primarily physical at all but rather psychological and characterological and, ultimately, narratological. Einstein offers to time travel writers a repertoire of new opportunities for experiments in storytelling, and, with this repertoire, the imprimatur of demonstrable physical rigor. With the increasingly well-established authority of the vocabulary of relativistic spacetime, if not, strictly speaking, with the mechanics and mathematical formalisms of relativity theory per se, time travel fiction discovers its most stimulating inducement to formal radicalism, indeed its primal modernist moment.

The Einstein Theory in Fiction: "The Man from the Atom"

We rejoin the history of time travel in 1923, when the sixteen-year-old fan and aspiring author G. Peyton Wertenbaker publishes a story called "The Man from the Atom" in a special "Scientific Fiction" issue of Hugo Gernsback's pulp radio magazine, *Science and Invention*.[16] Wertenbaker takes up, more or less wholesale, the melodramatic style of scientific romance usually described as "space opera." The protagonist, Kirby, has agreed to test a new invention of his friend, the iconoclastic genius Professor Martyn, a device that has the capacity to shrink or enlarge its user. The professor's scant description of this device is typical of Gernsbackian "scientifiction":[17] merely a small box strapped to the chest, with "six wires . . . which connect with the body"; when a small button is pressed, the machine immediately begins to alter the wearer's size.[18] Such a remarkable effect appears hardly to be constrained by principles of conservation of energy or mass. In passing, the professor explains that, in the machine's shrinking mode, "matter it removes from the body is reduced to a gaseous form, and left in the air," while in the growing mode, "the machine extracts atoms from the air which it converts, by a reverse method from the first, into atoms identical to certain others in the body."[19] Whatever its tenuous physical basis, Professor Martyn's box should ultimately allow Kirby either to "grow forever" until there is "nothing left in the universe to surpass," or, using the "reverse method," to "shrink so as to observe the minutest of atoms, standing upon it as you now stand upon the earth."[20]

Determined to use this "wonder machine" to enlarge himself despite "a multitude of unknown dangers," Kirby dons a space suit (a "sort of thermos bottle . . . with a transparent dome . . . made of strong unbreakable Bakelite"), presses the appropriate button, and begins to grow.[21] Soon, the Earth "became more and more like a little ball a few feet thick"; then "the stars were circling now about my legs"; and finally even whole galaxies, "the nebulae," shrink together into "a vast nucleus of glowing material."[22] Eventually, Kirby grows so colossal that he passes entirely beyond the bounds of his universe and into an "ultra-planetary space," a macrocosmos within which his own "nebula" is the size of only a single atom—hence "the man from the atom." Such a metauniversal viewpoint will eventually become common in time travel fictions, at times even an indispensible narratological medium, as I discuss in chapters to follow. In Wertenbaker's story, we may already note how easily an *outré* physical mechanism like a size-changing box can give rise to a peculiarly *narrative* setting, a transcendental quasi space in which entire "universes" are available to be viewed or reviewed, and the vast "empty temporality"[23] of whole cosmologies is distilled to a handful of sentences.

The story's opening lines have the reader encounter Kirby at the climactic moment in which he is already stranded in this macrocosmos, his own universe hopelessly lost among the innumerable "atoms":

> I am a lost soul, and I am homesick. Yes, homesick. Yet how vain is homesickness when one is without a home! I can but be sick for a home that has gone. For my home departed millions of years ago, and there is now not even a trace of its former existence.[24]

This soliloquy already contains hints of the Einstein theory, or at least of the tenor of its conceptual terms, for it turns out that space and time are closely linked for Wertenbaker, so that to be millions of times too large also means, somehow, to be "millions of years" too late. What Kirby and Professor Martyn have both failed to anticipate, in the new science of "atomic energy" that controls the size-changing box, is that

> time is relative, and depends upon size. The smaller a creature, the shorter its life. And yet, to itself, the fly that lives but a day has passed a lifetime of years. So it was here. Because I had grown large, centuries had become but moments to me. And the faster, the larger I grew, the swifter the years, the millions of years had rolled away.[25]

The passing comment that "time is relative," lent no more nuance than the mechanism of enlargement that first lands Kirby in this predicament, operates less as a theoretical claim than as cultural jargon, in somewhat the same manner in which a current science fiction author might mention in passing that a landscape is virtual, or that a ship is entering a wormhole. Wertenbaker's calculations of the quantity of spacetime traversed during the plot are, to say the least, haphazard: "millions of years," "millions of eons," "a trillion centuries," "billions of trillions of light-years." But the upshot is that a certain all-purpose aura of relativity theory allows Wertenbaker to connect, however loosely, changes in size, speed, space, and time, all within an assemblage of physical phenomena presented straightforwardly as scientifically realistic. The strict physics of the Einstein theory necessarily remain a black box, represented literally by the unexplained "box" attached to Kirby's chest. The theory's conceptual *effects*, however, therefore become all the more fodder for a speculative, rather than merely fantastic or fantasmatic, Gulliveresque travel tale, as well as for metanarrative reflection upon the scientific significance of such travel.

Thus Kirby's assertion that "because I had grown large, centuries had become but moments to me," and in general the notion that "time is relative and depends on size," are direct echoes of the Lorentz contraction and Minkowski's description of a spacetime manifold. Science fiction genre is still years away from the (ostensibly) more rigorous demands of "hard" writing that will be promulgated by John Campbell et al., and Wertenbaker clearly feels no especial obligation to chase down the inconveniently complex details of Einsteinian physics. When Kirby does ask Professor Martyn directly, "How does it work?" the professor replies: "[Y]ou could not understand all the technical details. It is horribly complicated."[26] Indeed, Martyn himself seems unfamiliar with the inner workings of his own technology:

> As I said, I have little idea of my invention except that it works by means of atomic energy. I was intending to make an atomic energy motor, when I observed certain parts to increase and diminish strangely in size. It was practically by blind instinct that I have worked the thing up.[27]

In Martyn's naïve but prodigious "blind instinct," we again see the character traits of the independent inventor/engineer à la Twain's Hank Morgan, Wells's Time Traveller, or Mackaye's Copernicus Droop, or perhaps an unselfconscious parody of Edison or Einstein himself—in any case, a fantastical figure of "heroic middle-class individualism" whose real social obsolescence will not be fully reflected within science fiction until long after Heinlein's or Asimov's rocket-building bootstrappers have themselves come to look like unintentional parodies.[28] Other remnants of older scientific discourses are also evident, and freely admixed with newer "atomic physics" in Wertenbaker's plot exposition: electricity, hydrodynamics, mechanics, biological evolution, and so on.[29] But a good part of the story's interest for its readership, first in *Science and Invention* and later in its reprints in *Amazing Stories* and beyond, is clearly in the explanations themselves, both narrative and scientific, of the extraordinary spatial and temporal changes undergone by the protagonist, as well as in the impression that such explanations are in agreement with up-to-date science. The macrologue has now come of age; no longer a preface or a frame, it is the heart of the story.[30]

Toward the end, with Kirby still stranded in his macrocosmos, his attention is understandably devoted to his own quandary—namely, the problem of how to return home to his "atomic" world—and he seeks a solution both scientifically and narratively plausible. Almost as an aside, Kirby finds himself residing in a futuristic society far advanced in both scientific and social evolution. In Wertenbaker's brief, almost fleeting description of this society, one spots the sociopolitical vestiges of utopian romance or its dystopian spinoffs, which lend this author a convenient, perhaps even default, language for sweeping transhistorical adventure. The following passage clearly invokes motifs from stories such as Wells's *The Time Machine* or Ignatius Donnelly's *Caesar's Column*, which had long since become canonical sources for far-future imagery:

> This strange planet of a strange star is all beyond my ken. The men are strange and their customs curious. Their language is beyond my every effort to comprehend, yet mine they know like a book. I find myself a savage, a creature to be treated with pity and contempt in a world too advanced even for his comprehension. Nothing here means anything to me.
>
> I live here in sufferance, as an ignorant African might have lived in an incomprehensible, to him, London. A strange creature, to play with and to be played with by children. A clown . . . a savage . . . ![31]

However, this brief description virtually exhausts the references to futuristic society, in keeping with the extremity of temporal dilation in which the story indulges overall. The generic shift from utopian romance to time travel is fully accomplished; the priorities of "The Man from the Atom," in terms of the attention it gives to the means rather than the ends of travel, directly reverse those of a book like *Looking Backward*, even as the elements of narrative and thematic structure (time travel, past and future romance, sociopolitical contrast) remain, in their schematic outlines, very similar.

Relativity and Return: "The Man from the Atom (Sequel)"

In 1926, Hugo Gernsback reprints Wertenbaker's tale in the inaugural issue of *Amazing Stories*. At the same time, he commissions from Wertenbaker a sequel for the next issue, and indicates in the text, by inserting the words "The End of Part One," that "The Man from the Atom" will be a serial. During the roughly three years that intervene between the writing of the two parts of Wertenbaker's story, the popularization of the language of atomic energy and relativity greatly expands, as does the extent to which the discipline of physics is thought of as the principal basis for the natural sciences. The traces of these scientific and cultural shifts, and of the pressures they place on the plausibility of "physical" science fiction, are easily detectable in Wertenbaker's sequel, which begins by explicitly couching the motifs of the first story more directly and firmly in the language of the new physics. This is a language of which pulp readers are by now well aware (even if largely by hearsay), and that is in itself sufficiently dramatic to alter the principles of plausibility, natural law, and even "fate" that Wertenbaker had clearly construed as the basis for his prior tale. Indeed, by 1926, the facts of physics have become so much stranger than fiction that Wertenbaker has little difficulty appending his own characteristically space-operatic style directly to his scientific expositions:

> I never hoped—never dreamed, when I wrote the tale you have read, that I should ever see the earth again. Who in the universe could have hoped against all the knowledge of insuperable fate which had come to me? Who could hope to overcome Time and Space, to recapture that which was gone forever? Yet it is just this that I have done—or something very like it. And it is a story a thousand times more fantastic, more impossible, than the story of my journey. And like that it is true.[32]

The plot of "The Man from the Atom (Sequel)" revolves around the formulation of a scientific, or at least scientific-sounding, procedure by which the hopelessly marooned Kirby might get himself home. Not coincidentally, the sequel also supplies a metanarrative explanation, more or less coherent, for its own existence as a "true" manuscript about these events, a quasi-narratological or paratextual report of how its narrator came to arrive back on his original Earth to publish the tale, very much like the epilogue of Bellamy's *Looking Backward* or the appended preface of Thomas's *The Crystal Button*. Kirby, briefly recalling his original enlargement and stranding, now sets out to depict more fully the society of the "planet of the star Delni" on which he finds himself, in part in order to develop the love story between himself and Vinda, "the daughter of the King of the planet," but also to provide expository grounds for a scientific account of his return to Earth. It turns out that "Vinda's father, the King, was a physicist," and generally, in Delnian society, as in the future societies of many of the utopian romances, the science of travel and the scientific foundation of future culture usefully collude.[33] Wertenbaker thus has Vinda and the king make direct references to their own understanding of the Einstein theory (as does Gernsback himself, incidentally, in his brief preface to Wertenbaker's sequel):[34]

> "You have spoken," [Vinda] said, "of a man called Einstein on your earth, and of other men who believe that time is a fourth dimension and that it is curved. Some of them, you say, believe that space is so curved that, if one goes sufficiently far, he will return to the point from which he started."[35]

Here Wertenbaker's sequel becomes more than just a continuation of the prior story. In the years since his debut in *Science and Invention*, Wertenbaker has reconsidered the scientific and narrative basis of his earlier work, and now weaves into the new plot a substantial revision of the natural philosophy underlying the narrative, fashioning current physics into a rationale for the reconciliation of what had seemed, according to the slightly older physics invoked by the first story, an impossible spatiotemporal and narratological quandary.

An incipient "hard" science fiction now imposes its putative rigor upon the fictional conventions of space opera. The exposition in Wertenbaker's sequel is elaborately macrological, far more so than the scant explanations of Professor Martyn's size-changing box in the first story; indeed it competes, in terms of both length and centrality, with the love story that occupies Kirby for the rest of the plot. What Wertenbaker presents is a full-scale philosophical treatment of temporality, in which "dimensions," "natural forces," "cycles of time," and similar terms ostensibly borrowed from the new physics are combined with a quasi-Nietzschean cyclical model of history that might easily have seemed, both to Wertenbaker and to his readers, to be a plausible corollary of relativity theory. "[O]ur evidence," Vinda tells Kirby, "has taught us that time goes in circles, in cycles. They say that, if one were to live forever, he would find eventually the whole of history repeating itself."[36] Because, as we know from the first story, Professor Martyn's size-changing box also accelerates or decelerates time, Kirby need only increase his size until he passes entirely out of this particular macrocosmic "cycle" and comes around again to his own historical moment. The expert scientists of Vinda's world calculate the precise parameters of the journey, and Kirby sets off by once again pressing the "enlarge" button. Becoming sufficiently gargantuan, he eventually returns to his own past/future at just the right size, arriving back in a new cyclical instantiation of twentieth-century America.

The heroic incoherence of this cosmological jaunt is no more of an impediment, in principle, than the blank inscrutability of the size-changing box in Wertenbaker's first story. However, even the three years separating the two pieces have considerably upped the formal and conceptual demands imposed upon the physical vocabulary of the story's macrological apparatus. Such demands augment, for instance, the degree to which the author directly indexes current science like the Einstein theory in order to satisfy the genre pressures of time travel fiction, a trend contemporaneously reinforced by Gernsback's own continual efforts to differentiate "scientifiction" from other subgenres of romance, adventure, and fantasy. One of Wertenbaker's central conceits in "The Man from the Atom (Sequel)"—that the men of Delni are scientifically advanced beyond Kirby's comprehension, while the women (including Vinda) "had never evolved . . . beyond a certain state of civilization," and can therefore empathize with his own "rudimentary" intellect[37]—provides him with a convenient means to incorporate the up-to-date terminology of physics without having to theorize it extensively, since neither Kirby nor Vinda is capable of comprehending the science behind the plot. Vinda declares:

> "[W]e are not able to understand it as the scientists do. They speak, for instance, of the dimension of *size*. It seems that there is a direction, which we cannot quite grasp mentally any more than we can grasp time as a direction,

which extends from the small to the great. That is to say, when you grow you are really moving in a new dimension which is linked, how I do not comprehend, with the dimension of time."[38]

"But really," Kirby agrees, "that is too obscure for me."[39] As in Mackaye's *The Panchronicon*, universal sexism, the tacit assumption that societies everywhere would naturally perpetuate a strictly gendered division of scientific labor, helps the reader to accept the limited exposition of relativity physics that Kirby and Wertenbaker are capable of providing.

Such vocabulary, as it were an elliptical or paraliptical relativity physics in the guise of conventional narrative's constraints on exposition, gains an ever-increasing foothold as the pulp "super science" magazines evolve from eclectic collections of fantasy, action-adventure, and the supernatural into more specialized niche venues for what we recognize, from the perspective of roughly post–World War II popular culture, as science fiction proper. As early as the mid-1920s, in the original stories being published in Gernsback's *Amazing Stories* and William Clayton's *Astounding Stories of Super-Science*, one encounters conceptual amalgamations or apothegms such as Einsteinian "time-space,"[40] relativistic (faster-than-light) travel,[41] "inter-dimensionality,"[42] "infra-dimensionality,"[43] "hyperspace,"[44] temporal "acceleration" or "deceleration,"[45] the conversion of mass into energy,[46] spacetime "curvature,"[47] multiple universes and "world lines,"[48] and, especially, the "fourth dimension."[49] By about 1938, when the editorship of *Astounding Stories* is taken over by John Campbell, nearly every issue contains at least one time travel story, and nearly all of these stories, as well as numerous nonfiction columns and readers' letters, explicitly refer—in their slightly oblique or self-obfuscated manner—to spacetime curvature, interdimensionality, and temporal relativity.[50]

More crucially, because of this trend toward a macrological physics, time travel, when it finally matures into a fully recognizable fictional category in the mid-1930s, becomes a genre inherently—one might even say, as odd as this sounds, organically—metanarrative, a genre *about* the basic conventions of story construction, just as detective fiction might be viewed as a genre inherently (or organically) about the reading and interpretation of signs. Indeed, time travel fiction is narratological in very much the same sense that detective fiction is semiological, by virtue of the formal and aesthetic demands built into the plot structures either provoked or constrained by its science, or, in other words, in the guise of an inherent and almost reflexive response to genre expectations. Moreover, the formal turn that characterizes time travel fiction after Einstein coincides nicely with Campbell's highly influential insistence on scientific plausibility and consistency—its chief result, finally, is the paradox story.

Thus we ought not to be surprised by a specific characteristic of 1930s time travel fiction—"paradox fiction" is possibly a better term for it—that could otherwise have seemed perplexing, given the close affinities of the genre with physics and relativity theory. The effects of Einsteinian physics on time travel fiction are *not* primarily physical at all, in the sense of a set of themes or topics that become fodder for authors' or characters' speculations. Rather, as I suggested earlier, the effects are initially psychological and then predominantly narratological, in the most comprehensive sense of this latter term. What physics finally enables within science fiction is a metaliterature of Oedipus and Narcissus, a literature about encountering (or reencountering) oneself, about meeting (or remeeting) one's progenitors, about negotiating (or renegotiating) one's personal and historical origins. It takes many years for time travel writers to discover the psychological implications of the narrative possibilities opened by the new physics, but once they do, these *topoi* effectively conquer the thematic terrain of time travel fiction as effectively as the metaphorics of Darwinistic evolution had done within utopian romance a generation earlier. For the rest of its history as a narratological laboratory—arguably even into the contemporary period, post–Stephen Hawking, dominated by new "hard" speculations about the quantum structure of spacetime or about the multiverse—time travel fiction is a genre of psychological implication, a scenography in which selves meet themselves, kill their progenitors, and plumb the significance of their own histories for their present instantiations or avatars. Time travel, in essence, becomes what Lacan thought the psychoanalytic session was, a "realization of the [subject's] history" in a present discourse, or even "the restitution of the subject's wholeness . . . in the guise of a restoration of the past."[51] In my final chapter I discuss some of the most extreme theoretical implications to which such a psychological/narratological bias may commit this fiction, or may drive its theoretically minded reader.

However, we may briefly observe the early traces of this psychological/ narratological turn in Wertenbaker's "The Man from the Atom (Sequel)," where their complicity with the restructuring of spacetime in Einsteinian physics emerges through a series of brief allusions to historical paradox and psychological duplication. Soon enough, such allusions will become a central obsession of time travel fiction, but for Wertenbaker they are merely curiosities or artifacts of the type of plot his physical terminology has opened up. "Is it true," Kirby asks, late in his adventure, "that there will be another incarnation of my body which will leave the earth at the same time I am returning?"[52] As it turns out, there won't be such an incarnation in this particular story, although near its end Wertenbaker speculates briefly that "the Kirby who left the world of this cycle is not the Kirby who has returned," and that there is "another person, my double in appearance, life, and name, who is now wandering about the universe, watching with amazement the formation of the stars."[53] Of course, narrative literature has often dabbled in such duplication as a problem of form (and its philosophical entailments); a classic instance, following Genette's extended analysis of Proust's monumental self-reconstruction in *Remembrance of Things Past*, is the "double narrative" of Marcel, which continually juxtaposes earlier and later modes of the character's self-understanding, "modif[ying] the meaning of past occurrences after the event."[54] A certain similar innovation is concocted in these early pulp stories around the directness with which the presence of the double, or the reiteration of histories, is mentioned. Within post-Einsteinian time travel fiction, the double can appear for the first time as a literal character on the level of immediate plot, and history can be altered, not as a matter either of speculative fantasy, sociopolitical counterfactuality, or psychological allegory—say, for the purposes of utopian critique or autobiographical reflection—but rather purely on the level of story itself, as an actual event within a narratively realistic *mise-en-scène*.

Hence Wertenbaker's almost passing comments about the "other" Kirby are as prescient as Kirby's brief inquiry into the specter of historical revisionism summoned by relativity theory: "In all theories of time as a dimension, this point has always raised itself in my mind. If I were to return during some crisis in history and foretell the mistake that would be made, could that mistake be rectified, changing the whole course of history?"[55] The answer here again is no—Kirby's return could not change history—a conservative or conservationist answer that will recur often in subsequent time travel fiction: the historical past is figured by Wertenbaker strictly as "inevitable" or as a "destiny." However, oddly and fruitfully, Wertenbaker is not loyal to his own conclusion about such historical inevitability, and when Kirby arrives back in his own time, he discovers some odd revisions:

> The world has changed in many details since I knew it in the last cycle. For instance, the America I knew was a Republic still, whereas now, you know that it is the Monarchy which was declared by Theodore Roosevelt during the Great War of 1812, and which is now ruled by the Emperor Theodore II.[56]

The two themes evoked in this passage, doubling and historical revisionism, are grounded in what might seem to be a natural inference from the complex and counterintuitive deconstruction of simultaneity with which Einstein's theories first imbue the popular imagination. For Wertenbaker, they are still sidelines to the more central task of expositing a realistic mechanism for intracosmic travel. Yet the themes prefigure what will soon become the most abiding concerns of time travel plots, as the formal and psychological questions of revision, self-conception, and causal conservatism are driven to the foreground.

By the late 1930s and early 1940s, the popular understanding and acceptance of relativity theory have advanced considerably, and the psychological and epistemological extremities at which Wertenbaker's story only hints move directly to the surface; the time travel story now begins to attain its "paradox" mode. This is certainly the most familiar mode or pattern of time travel fiction for readers of science fiction today, not to mention for viewers of time travel narrative in film and on television. The essence of a paradox story is interference in the past and the potential irreconcilability of lines of events that such interference sets up. For instance, my action as a time traveler might preclude my setting off on the very voyage by which I take that action; or, as in Gregory Benford's *Timescape*, some urgent message I send to the past might alter events such that the very sending of the message becomes either impossible or superfluous. There are numerous subtypes of the paradox story, each with varying psychological and historical entailments, but in all cases, what these stories create is a powerfully evocative epistemological conundrum, a situation in which questions of the world that is known, and of the figure by whom it is known, are rendered equivocal or even fundamentally undecidable. No doubt, literature has always been able to reflect upon its own capacity to detach the question of narrative viewpoint from the process of storytelling, and thereby to put into speculative play the location or position of the reader of narratives—the doubled character is one technique for doing so, the juxtaposition of multiplied viewpoints another. In the time travel paradox story, such an epistemological condition is literalized, and its conceptual imposition on the reader therefore made palpable in a way substantially different from that of experimental literature or metanarrative. Let us see how this happens.

Relative Misrecognition: "By His Bootstraps"

My next example, Robert Heinlein's first time travel paradox story, "By His Bootstraps," is published in *Astounding Stories* in October 1941 under the pseudonym Anson MacDonald. This story establishes a number of parameters for the ways in which science fiction writers, during the years of John Campbell's editorial tenure and after, will treat time travel and the philosophical problems it raises. The story also helps cement several distinctive Heinleinian modes: a sardonic, even self-satisfied sexism, a glib militarism bordering on crypto-fascism, and a nonchalant inheritance of a Hank Morgan–esque libertarian individualism (alluded to in the story's title) that will be enormously influential, at times even *de rigueur*, within the genre of "hard" science fiction that Heinlein, along with several other young writers introduced or promoted by Campbell, helps to establish. However, most important, and despite its sociopolitical indulgences or blindnesses, "By His Bootstraps" attempts to confront head-on its own considerable psychological and narratological complexity, enabled by its assimilation of what it takes to be current physics and canonical philosophies of mind.

The plot of "By His Bootstraps" depicts the adventures of Bob Wilson, who travels through time and encounters himself at various moments.[57] As the story opens, Bob is secluded in a locked room, frantically working to complete a thesis entitled "An Investigation into Certain Mathematical Aspects of a Rigor of Metaphysics."[58] Typical of Heinlein, the ironic futility of such a purely academic topic is quickly made obvious: this "metaphysic[al] rigor" will forthwith become an entirely physical problem for Bob, at the same time as the locked-room mystery that the story sets up will be lent a satirical science-fictional twist. As Bob continues to type away, absorbed in his work, a second character steps into the room through what we will soon discover is a "time gate," essentially a vertical hole in midair. This intruder directly informs Bob of what the reader already suspects from Heinlein's acerbic tone, that Bob's thesis on the metaphysics of time travel "is a lot of utter hogwash."[59]

Following a brief and slightly uncanny conversation, the intruder tries to persuade Bob to go back with him through the gate. For the moment, the new character is identified only as "Joe," but as the reader very soon realizes, Joe is in fact a later Bob Wilson—I will identify him as "Bob 2"—whom Bob 1 will eventually become as the plot proceeds. Soon after, a Bob 3 also appears in the room, debating with Bob 2 whether Bob 1 should stay put instead of venturing through the time gate. Even a Bob 4 briefly calls on the phone, apparently to check on the progress of the other Bobs as they age into one another. Eventually, Bob 1 gets shoved through the time gate, where he encounters his ostensibly final avatar (call him Bob 5), an older man who goes by the alias Diktor. From here, the plot is arranged to untangle both the identity problem and the locked-room mystery, so that the reader eventually discovers how the original Bob 1 can become, in a consistent succession, the other four Bobs, while at the same time finding out by what means Bobs 2 and 3 get into Bob 1's room. From the metaviewpoint we eventually train ourselves to adopt, the unraveled or reconstructed plot runs something like the accompanying figure I have drawn. Diktor, having figured out the schema of his own self-fulfilling origin, arranges to send the newly arrived Bob 1 back (as Bob 2) to persuade "himself" to come through the gate. Bobs 3 and 4 interrupt Bob 2 on this mission to corral Bob 1, and there are incidental complications, such as Bob's divesting himself of a typically Heinleinian shrewish girlfriend, and a brief sortie to acquire books and materials to assist Bob/Diktor's future ascendancy on the other side of the time gate. But finally, the evolving Bob becomes Diktor, and is "now" in a position to arrange the whole sequence. The story concludes in properly looped fashion, as Diktor prepares to send Bob 1 back through the gate (as Bob 2) into the locked room of the opening scene.

Diktor, as it turns out, has considerable incentive to initiate this sequence of encounters. He is the ruler of an entire future world, served by a

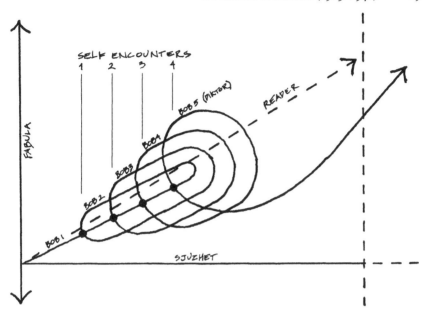

"By His Bootstraps," plot diagram

harem of nubile Weena-like slaves, operating with the impunity of Heinleinian heterosexual male birthright. Indeed, Diktor even instructs one of
his earlier Bobs to import political tracts and self-help books to aid him in
constructing his paternalistic autocracy.[60] Here the whole future, its pas-
tures dotted with advanced but dilapidated technology, its degenerate but
concupiscent humans primed for governance and husbandry by Yankee
savoir faire, reads like a poor parody of Wells's dystopian pastoral, as if we
are provisioned with Wellsian sex and good food yet have fully—that is,
fully *successfully*—repressed the Morlocks who must produce or sustain
these commodities. In essence, Heinlein indulges what John Rieder calls
the "anthropologist's fantasy" underlying the "colonial ideology" of such
science fiction: "The thrill of the technological breakthrough is not that it
benefits everyone but that it produces a singular, drastic difference between
those who possess the new technology or power source and those who
do not."[61] Futuristic fascism is depicted as *de facto* scenography, and thus the
sociopolitical elements of Wells's *The Time Machine*, which remains the
most obvious source for descriptions of such a far-future society, become,

so to speak, painted background, inconspicuous by virtue of their mere unelaborated presence.[62]

In effect, by taking them utterly for granted, Heinlein has sacrificed such sociopolitics for (or submerged them within) a problem that is closer to home, as well as better suited to the extreme libertarian ethos that allows a potentially utopian/dystopian plot to hinge so singularly on the wiles of its lone protagonist: what happens when Bob Wilson meets himself? Time travel is here a question not about the future or the past but about the present and, more specifically, the *narrative* present: what happens when you encounter "another" you—and who then (or which) is "yourself"?[63] Utopian prognostication is displaced by psychology and, in turn, by epistemology; but epistemology, in its own turn, becomes a problem of narratability. It is vital to the progress of "By His Bootstraps" that there be, from the very beginning, an especially prescient viewpoint available to the reader to connect Bobs 1 through 5, all of whom are themselves relatively slow on the uptake. The reader figures out, long before Bob, that there is only one character in this story (discounting Diktor's anonymous slaves and Bob's swiftly discarded girlfriend), although Bob, as Diktor, eventually comes around to this knowledge in the story's conclusion. Here, for instance, is Bob 2's first encounter with Bob 1, as he (Bob 2) returns through the time gate and into the plot's second iteration of the locked room:

> Bob Wilson stepped through the locus known as the Time Gate. . . .
>   There was a man in [his room], seated at his own desk. . . .
>   He felt a passing resentment at finding someone at *his* desk in *his* room. . . .
>   The chap did look vaguely familiar, although all he could see was his back. Who was it?[64]

At moments of the story, Bob will recognize himself, at other moments not. Heinlein goes to some length, and to some expense of plausibility, to emphasize the difference between Bobs in various self-encounters. The otherness of the self in these meetings is a question not just of mere identification, but rather of character or personality that diverges from self-image. A short time later in the plot, Bob 2 "suddenly wondered if the other man *could* be himself. The stupid arrogant dogmatism of the man's manner infuriated him."[65] Of course, such misrecognition is necessary, if for nothing else at least to keep Bob from gleaning too much information from his other selves and halting the plot in its tracks. We can note here that what had begun ostensibly with physical theory, or at least with "a rigor of metaphysics," very quickly comes to resemble something like Lacanian psychoanalysis: the subject is *required* to misrecognize himself. Why? To become what he eventually will have been (Diktor)? To make it possible for the narrative of his subjectivity to proceed (he must *become* himself, therefore not already *be* himself)? Or to demonstrate that these two eventualities— becoming oneself, becoming one's story—are ultimately the same thing, and that both entail not being (oneself, one's story) yet? Slavoj Žižek describes this exigency as "the positivity proper to the misrecognition,"[66] and, referring to Heinlein's later time travel novel *The Door into Summer*, explicates the "basic paradox" of Einsteinian time travel in psychoanalytic terms:

> This, therefore, is the basic paradox we are aiming at: the subject is confronted with a scene from the past that he wants to change, to meddle with, to intervene in; he takes a journey into the past, intervenes in the scene, and it is not that he "cannot change anything"—quite the contrary, only through his intervention does the scene from the past *become what it always was*: his intervention was from the beginning comprised, included.[67]

Just so, Bob evolves into Diktor, into himself—not despite his own interference with his past but because of it. At the same time, the story becomes itself, resolves itself, comes around to conclude itself, because Bob was not yet himself (Diktor) and had to become that—essentially, he had to produce the *narrative* that led to himself. That narrative proceeds only because of its own series of missteps or misrecognitions, without which the plot would precipitously end and Bob fail to arrive at his completed (or, in terms of the story's overt metaphysics, causally determined) final avatar, Diktor. Here is the first dysfunctional meeting between Bob Wilson and himself:

> He saw a chap about the same size as himself and much the same age—perhaps a little older, though a three-day growth of beard may have accounted for that impression. . . . Wilson decided he did not like the chap's face. Still, there was something familiar about the face; he felt that he should have recognized it, that he had seen it many times before under different circumstances.[68]

And here is another, in which Bob 2 is surprised to discover after the fact that he is (or has been) "Joe":

> He felt suddenly startled at his own words. It was at this point that he realized that he was *in fact*, "Joe," the Joe whom he had encountered once before. . . . Hearing himself refer to himself as Joe slapped him in the face with the realization that this was not simply a similar scene, but the *same* scene he had lived through once before—save that he was living through it from a different viewpoint.[69]

Bob, then, is required to work out, although not precisely in these terms, the basic psychoanalytic problem that Lacan constructs and Žižek glosses: that the self *is* what it *will have been*, but necessarily does not realize it in the process: "This introduces a relationship between truth and misrecognition/misapprehension by which the Truth, literally, arises from misrecognition."[70] Self-identity is never that which is felt or experienced *in situ*, but rather that which is first mis-taken, then dis-covered, and ultimately re-narrated from a different viewpoint.

Heinlein speculates upon this fundamental problem of divergent (narrative) self-identity, but he deliberately has his protagonist, Bob, do so through a quasi-Cartesian methodology that in the end will not suffice to explain the temporality of what here gets called the "ego":

> The ego was himself. Self is self, an unproved and unprovable first statement, directly experienced. What, then, of the other two? Surely they had been equally sure of ego-being—he remembered it. He thought of a way to state it: Ego is the point of consciousness, the latest term in a continuously expanding series along the line of memory duration.[71]

This seems right to Bob, so long as "memory duration" continues to correspond to world duration—that is, so long as only one "point of consciousness" is present at any one time, and therefore the ego *through* time remains the same one that appears *at* any given point of time. Of course, this is precisely what does not happen in the time travel paradox story, where the subjective sense of linear progress conflicts with, or crosses, the objective sense of linear history. The ego, as a point of consciousness in a "continuously expanding series," can remain a plausible fiction only so long as the memory line does not double back, does not return to or repeat its own prior line of duration.

The "moving point" model of consciousness that Bob attributes to Descartes is more immediately traceable to Heinlein's own gleanings from contemporary physics, in this case the concept of a worldline, a path through four-dimensional spacetime.[72] This is a physical concept that Heinlein had already retrofitted for a theory of selfhood a couple of years earlier, in the story "Life-Line":

> "[Y]ou are a space-time event having duration four ways. . . . Imagine this space-time event . . . as a long pink worm, continuous through the years, one end at his mother's womb, the other at the grave. It stretches past us here and the cross-section we see appears as a single discrete body."[73]

One can already note that for Heinlein the immanent perspective that perceives the self as a mere "discrete body"—in Bob's quasi-Cartesian terms, a "point of consciousness"—is in "Life-Line" explicitly declared to be an "illusion."[74] The model of a line, once posited, immediately suggests and demands a transcendent viewpoint, in essence a historical overview that tracks its own progress from above. A full self, one with a whole history of resynthesized "discrete" moments, is properly perceivable only from outside the four-dimensional "pink worm" of its progress through spacetime, and the "point of consciousness" is thus redescribed as a mere "cross-section" of the worm. In fact, the cross-section, taken solely on its own, is that same ostensible Cartesian ego Bob initially proposes in "By His Bootstraps," an "unproved and unprovable first statement, directly experienced." Such a pseudoself will continually misrecognize itself in any other time or place, and must see its own previous or subsequent spacetime cross-sections only as others, never as itself. Because "Life-Line" is not a time travel story per se, we don't get to witness this misrecognition directly in the plot, as we do in "By His Bootstraps." However, "Life-Line" at least does start to depict for us the point of view from which such misrecognition, as well as its ultimate recuperation—the completed Diktor-like overview of the narrative of life's "pink worm"—would occur. I will have considerably more to say about such a point of view, especially in my final two chapters.

In "Life-Line," the retrospective overview one needs, first prompted by the terminology of physics but adapted to psychological theory, is furnished by a machine invented by the protagonist, Dr. Pinero; it is a "mass of equipment" that "vaguely resembled a medico's office x-ray gear."[75] As with the size-changing box in Wertenbaker's "The Man from the Atom," the mechanics of Pinero's device remain obscure to the story's other characters, and hence to the reader: "Beyond the obvious fact that it used electrical power, and that some of the dials were calibrated in familiar terms, a casual inspection gave no clue to its actual use."[76] But the physical operation of the machine is far less important to the story than the crucial aspect of psychological and narratological theory it literalizes. Pinero uses his machine to predict dates of death, allowing his paying clients to bilk their life insurance companies by buying policies at opportune moments. Effectively, what the machine does is record a *story* about the whole course of an individual's life, a fully comprehensive narrative report, in lieu of, for instance, the more vague actuarial data used by insurance company statisticians. Continuing to mix metaphors—"line," "worm," "vine," "cable," "corridor"—but in all cases explicating the human self as a linear thing viewed from a nonlinear external perspective, Dr. Pinero describes what amounts to a machinery of narrative closure:

> "I asked you to think of life as electrical. Now think of our long pink worm as a conductor of electricity. You have heard, perhaps, of the fact that electrical engineers can, by certain measurements, predict the exact location of a break in a trans-Atlantic cable without ever leaving the shore. I do the same with our pink worms. By applying my instruments to the cross-section here in this room I can tell where the break occurs, that is to say, when death takes place."[77]

Thus the machine completes the story of one's life by extrapolating a consistent conclusion from the conditions furnished by the cross-section of the present. In essence, it re-cognizes the full series of spatiotemporal differences that constitutes a person's story, and of which specific "points of consciousness" are only misrecognized parts.

In "Life-Line," Einsteinian spacetime physics undergirds the psychology of transtemporal selfhood, a psychology that in turn is reified as a series of mechanical and electrical instruments that provide the narratological viewpoint required to project the closure of a fantasmatically complete self. "'You have . . . been told that time is a fourth dimension,'" Pinero begins, with his familiar terminology of physics, but he ends with the jargon of melodrama: "Pinero shook his head sadly. 'I am sorry for you, my dear Luke. You will die before tomorrow.'"[78] Luke, like Bob, is merely a "Cartesian" self *in situ*, having not yet acquired the transtemporal viewpoint necessary to recognize himself, his true and complete temporal series. Pinero's machine does it for him, as it does for Pinero himself by the narrative's end.[79]

If "Life-Line" starts to literalize the viewpoint one would need to complete the story or fantasy of oneself, "By His Bootstraps" further depicts the synthesis of cross-sectional and metaviewpoints required to construct any (story of a) self. We can observe very directly the advantage the time travel paradox structure offers over the merely expository theorizations of "Life-Line." The "time gate" and the "Einstein–Minkowski"[80] geography of spacetime that it introduces permit Bob literally to traverse the narratological terrain of his own history as one might peruse a four-dimensional diagram, or perhaps turn the crank of a peepshow. Here Bob 4, interloping in Diktor's palace, manipulates the time gate:

> He found the controls set at zero; making use of the same process he had used once before, he ran the scene in the speculum forward through ten years, then cautiously hunted with the space controls on zero. . . .
>
> . . . He would have given up, was about to give up, when, purely by accident, one more fruitless scanning happened to terminate with a figure in the field.
>
> It was himself, carrying two suitcases. . . . He recalled the situation when he saw it portrayed in the speculum. It was Bob Wilson number three, about to quarrel with Diktor and make his escape back to the twentieth century.[81]

Although not yet Diktor—which is to say, not yet having comprised the full series of his spatiotemporal selves—Bob now begins to construct a more adequate theory of selfhood, contradicting his earlier quasi-Cartesian model. He again asks, "which one was *himself*?" But this time the answer is explicitly anti-Cartesian; indeed, if Heinlein had been versed more fully in modern philosophy, he might have linked his speculations not to Kant and Descartes but to Hegel, Bergson, or Proust, and ultimately to Leibniz's *Monadology*—or perhaps he might have extrapolated an entirely new ego psychology directly from Einstein, based on the nonsimultaneity of events viewed from different coordinate frames:

> By sheer necessity he was forced to expand the principle of non-identity—"Nothing is identical with anything else not even with itself"—to include the ego. In a four-dimensional continuum each event is an absolute individual, it

has its space co-ordinates and its date. The Bob Wilson he was right now was *not* the Bob Wilson he had been ten minutes ago. Each was a discrete section of a four-dimensional process.[82]

This seems to be an improvement, in its explicit anti-Cartesianism, but is still not entirely adequate. Each four-dimensional "event," including each Bob at each moment, is an "absolute individual," a "discrete section," essentially a *monad*. But of course Bob himself continues to narrate this continual radical difference and, more important to the time travel story, continues to *experience* it as literal otherness, something to be directly viewed on the screen of a mechanical device. In short, Bob has yet to account for his own "Diktatorial" role as the overseer of the series, and so Žižek's Lacanian description still applies: "The initial 'illusion' of the subject consists in simply forgetting to include in the scene his own act—that is, to overlook how 'it counts, it is counted, and the one who counts is already included in the account.'"[83] Placed back into Bob's quasi-Cartesian terminology: self is *not* self but rather a *story* of self, a narrative progression toward complete self-identity, still necessarily partial and unrecognizable along the way, and only ever identical with itself (paradoxically) *after* the fact, when the unfolding series of misrecognitions and uncanny repetitions that made it into a story at all acquires the impression or illusion of wholeness and presence—"a complete transcript," as Bob Wilson glibly phrases it.[84]

Of course, the reader realizes already what the "complete transcript" will say—namely, that Bob is in fact Diktor and no one else. The plot now awaits Bob's own halting realization, his maturation into the completed self that the line of his own narrative implies:

> He was Diktor. He was *the* Diktor. He was *the only* Diktor. . . .
> In review, it seemed obvious that he must be Diktor; there were so many bits of evidence pointing to it. And yet it had not been obvious.[85]

Only now does Bob become aware that he has been reading his own plot in the "speculum" of the time gate controls, the crossings and re-cognitions of the line of his own past, continually re-traversed by the successive Bobs through the gate. The realization has two components: first, the anti-Cartesian insight that the self is not an object (a cross-section) but rather a *history*, a narrative; and, second, the narratological insight that the self is a *reader*, and that what it does in realizing itself is to catch up with the perspective of the actual or potential audience of its own story. Thus, whereas Dr. Pinero's death coincides with the conclusion of "Life-Line," Bob's realization coincides with the ostensible (re)commencement of the plot of "By His Bootstraps," as Diktor finally reopens the time gate in Bob 1's room and sends Bob 2 back through it. To combine these two insights: the self is the narrative of its own time travel, a fantasmatic invention of a mechanism by which it completes an excursion into its own past, and therefore the possibility—literal in a time travel story, presumably fantasmatic in real life—of a consummate viewpoint upon its full series of cross-sections.

Thus, of course, the self and the story do conclude at the same time, even, or especially, where the story appears to recommence in a loop. In a real way, self and story *are the same thing*, a convolution I discuss most fully in my final chapter on the paradox plot of the film *Back to the Future*. But what we may notice already in Heinlein is how, in the coincidence of story with self that ends his tale, the reader's viewpoint also comes to coincide with that of Bob/Diktor. This triple coincidence of story, self, and reader in turn suggests that time travel paradox fiction enacts a fundamental means by which a self constitutes itself through self-relating or self-viewing—its full viewpoint upon its own history, constructed via narrative:

> One is therefore tempted to see in the "time paradox" of science-fiction novels a kind of hallucinatory "apparition in the real" . . . : a circular movement, a kind of snare where we can progress only in such a manner that we "overtake" ourselves in the transference, to find ourselves later at a point at which we have already been.[86]

But it is Bob's continuing failure to *be* a comprehensive re-viewer, to "overtake" himself fully, that eventually allows him to succeed in completing the looped structure of the story—becoming Diktor, meeting himself as Bob 2, etc.—in short, that allows him to become precisely himself, and the story to become his story:

> [T]his superfluous detour, this supplementary snare of overtaking ourselves ("voyage into the future") and then reversing the time direction ("voyage into the past"), is not just a subjective illusion/perception of an objective process taking place in so-called reality independently of these illusions. That supplementary snare is, rather, an internal condition, an internal constituent of the so-called "objective" process itself: only through this additional detour

does that past itself, the "objective" state of things, become retroactively what it always was.[87]

I noted earlier that the reader seems always to be one or two steps ahead of Bob in this process of (narrative) self-recognition. Given the structure of deferred self-identification that Heinlein's story is beginning to construct, I could state this another way: the reader is ahead of the game in recognizing the sameness-in-difference that constitutes Bob-Wilson-the-*narrative*; hence, on my earlier illustration (page 69), I have inserted an oblique dotted line for the reader, who is already piecing together the hypothetical narrative line that connects the various Bobs directly. Now, this readerly precognizance is no artifact of just this time travel story; it is a structural necessity of any such story, and indeed of any story at all in which identity over time becomes questionable, which is to say (à la Heinlein's four-dimensional psychology), any story of a self. One need only imagine the degree of incoherence such a story would provoke if there were no structural position available from which someone could recognize that the five characters were all a single Bob, that all the different "absolute individuals" of spacetime were finally one self.

No, the extradimensional viewpoint from which we recognize that Bob 1 *is* Bob 2, *is* Bob 3, on up to Diktor, is a structurally necessary position, one that the reader occupies fairly quickly, and Bob less quickly, but that in the end is absolutely required of the peculiar form of fantasmatic narrative closure that is a human self. And from this viewpoint, the self is, of course, never a point on a continuum—that could be only the merest hint of a purely aesthetic consciousness, mere immediate perception or sensation. Rather, the self is a movement, a traveling back along a continuum, as well as a capacity to step out of that continuum, to review it, to tie its points together, to synthesize it. Bob, like all selves, must come to narrate himself in order to be himself. He must finally become the story about himself, an array of narrative tracks re-viewed from a perspective outside, and one that is able to go both forward and backward and to weave these directions together into yet a new track. Selves are stories—time travel stories. This is what Einsteinian physics allows Heinlein to begin to think, and what time travel fiction continues to allow us to speculate.

Three Phases of Time Travel / *The Time Machine*

> "History splits into two when you perform such an experiment. In the other history, there is another you—. . . . there is an infinite number of 'you's' [*sic*]—propagating like rabbits at every moment. . . ."
>
> "What an appalling thought," I said. "I thought two were more than enough."
>
> —STEVEN BAXTER, *The Time Ships*

Three Phases of Time Travel

In my first "Historical Interval," I proposed Harold Steele Mackaye's *The Panchronicon* as the first time travel story—maybe a perverse choice, yet compelled by the peculiar, halting origins of time travel fiction. Mackaye's novel arrives with the decline of a particular narrative mode I named the macrologue, a conglomeration of framing devices in utopian romance through which writers negotiate an uneasy compromise between the pressures of Darwinist evolution and their own ambitions toward realistic sociopolitical prediction. When the time travel story first breaks out of its ancillary role as utopian macrologue, it cannot help but seem belated or even jaded, a mishmash of dubiously convenient tropes contaminated by their links to the perceived grandiloquence or naïveté of utopianism, as well as of the diverse and zealous reactions against it. Indeed, frame narrators in the last Bellamistic romances—for instance, the bizarre "Voice" of Wells's *A*

Modern Utopia, discussed at the end of Chapter 1—exhibit a prolific discomfort with prior utopian seriousness, and tend to fashion in its place a stance closely resembling parody, a "narrative . . . presented as only narrative, as its own reality—that is, as artifice."[1] *The Panchronicon* is among the earliest works in which such discomfort with utopian macrologia congeals into an autonomous type.

In the 1920s, a second phase of time travel fiction commences, largely also as a compromise between plausible realism and popularized science. With the rise of relativity theory in the 1920s and with the increasingly widespread identification of physics as a definitive ground for the natural sciences, time travel fiction is lent a new conceptual basis, as well as a variety of newly legitimized plot tricks. I have identified this second phase of time travel as that of the paradox story, and suggested that it culminates in the early 1940s with a spate of "closed loop" narratives epitomized by Heinlein's "By His Bootstraps." Of course, possibilities for paradoxical loops are already implied in time travel fiction considerably earlier, as I briefly suggested with regard to the multiplied lines of G. Peyton Wertenbaker's "The Man from the Atom" stories. It may seem surprising that explicit paradoxes do not crop up until much later—indeed, for the most part not until the mid-1930s.[2] Either it does not cross the minds of science fiction authors or editors to lean on the paradoxical latencies of narrative doubling—the possibility, for instance, of multiple incompatible pasts, or of self-bilking—or else the generic constraints upon narrative plausibility in "scientific fiction" operate to preclude such plots. As late as 1937, Lester del Rey remarks that John Campbell of *Astounding Stories* rejects his story "The Faithful" specifically because it entails paradox: "Campbell said it was well written, but he wasn't interested in stories that went around in a circle of time."[3] What we may say with confidence is that only at a specific historical juncture does it become possible, perhaps even compulsory, for authors and readers to consider time traveling a means for potential illogic, and thereafter as an opportunity to reflect, not only on the coherency of causality or chronology, but on the possibility of their spectacular or absurd failure.

Considered from the perspective of these first two phases, the macrologue and the paradox story, the genre of time travel fiction achieves a certain apotheosis with the time loop story. Once furnished with the notion or the leeway (or both) to represent not just temporal dilation, compression, and reversal but also overlapping, doubling, contradiction, and causal irreconcilability—in short, the whole catalog of possible misrelations between plot events—time travel writers make paradox the principal motif and rationale of their stories. The *telos* of this trend—now construing time travel as a special form of realistic epistemological and narratological self-depiction—is then the self-canceling or self-creating paradox, in which an achronic or acausal anomaly is recast as a realistic (if not always fully logical or strictly physical) single closed plot line.[4] "By His Bootstraps" is such a story, of course, as are Eando Binder's "The Time Cheaters" (1940), Ross Rocklynne's "Time Wants a Skeleton" (1941), James Blish's "Weapon Out of Time" (1941), and numerous others from the early to mid-1940s. Stories such as these fulfill the narratological aims that time travel fiction tends to set for itself once it fully embarks on the transition from subsidiary macrologue to self-reflective causal paradox. It may be feasible to think of the early 1940s as in some sense the genre's intrinsic culmination.

The third phase of time travel fiction, from the mid–twentieth century to the present, is somewhat more difficult to delineate, in part because of the variety of its inventions and aspirations, and in part because it continues to evolve, both in its stylistic diversity and in the unprecedented range of scientific means it discovers at its disposal. Let me suggest only briefly the outlines of such a third phase—the remainder of this book will be devoted to analyzing some of its specific productions.

A considerable variety of closed loop stories follows the epoch of "By His Bootstraps." However, presumably because there are limits to the novelty of the "cunning literary move[s]"[5] available to solve the logical riddles such narratives pose, later loop stories pursue directions largely away from purely closed plots, and for a variety of purposes: as a prompt to reflect on existential or ethical choices (Robert Charles Wilson's *The Chronoliths*, the films *Groundhog Day* and *Primer*), as a framework for structuring large-scale epics or future histories (C. L. Moore's "The Man Who Walked Home," Stephen Baxter's *Manifold* trilogy, Greg Bear's *City at the End of Time*), as a modulated repetition of slices of time prompting strategic or moral reevaluations (Hiroshi Sakurazaka's *All You Need Is Kill*, Ken Grimwood's *Replay*, the film *Triangle*), as an exploration of melancholia or nostalgia (Philip K. Dick's "A Little Something for Us Tempunauts," Ursula Le Guin's "Another Story, or, A Fisherman of the Inland Sea," the film *Frequency*), as an impetus or subplot within broader adventures or dramas (the television show *Lost*, the films *Il Mare* and *Terminator II: Judgment Day*), or as a parody of the arbitrariness of narrative choices or generic coincidence (David Gerrold's *The Man Who Folded Himself*, *Bill and Ted's Excellent Adventure*, numerous episodes of *Doctor Who*, *Red Dwarf*, and *Futurama*).

Above all, however, the third phase of time travel fiction is characterized by the ongoing *visualization* of parallel or multiplied lines of narrative, and with the aesthetic or technological depiction of such multiplicities. Noticeably, later paradox stories are far more likely to resolve themselves in altered or alternative worlds—again, often regardless of logical or physical consistency—than to close themselves strictly in single loops, the way that "By His Bootstraps" does. In one sense, this trend is due to advances in the physics adopted by time travel writers to undergird paradox stories, specifically Everett's "many-worlds" interpretation of quantum mechanics and its successors, which I will discuss further in Chapter 3, as well as a variety of theories and popularizations of multiverse physics. But in another sense, the depiction of spatiotemporal alternates, both physical and counterfactual, emerges from an inherently visual bias within time travel narrative itself, possibly even an inherently *cinematic* bias, which serves to make film, in a strong sense, the genre's principal or primal medium.

The third phase of time travel fiction, which is still developing, might simply be called "post-paradox," but I prefer to call it the "multiverse/filmic" phase. This clunky designation at least helps to characterize the diversity of story types that now occupies the category of time travel, and the futility of isolating either a single clear-cut set of generic features or a coherent historical trajectory for this fiction, as I was perhaps more readily able to do for its first two phases. Nonetheless, in its "multiverse/filmic" phase, time travel becomes all the more potent a resource for narratological inquiry. In the following chapters, therefore, I proceed less strictly with regard to definite historical or epochal moments, and instead examine a series of individual case studies, especially as they tend toward visual depictions of time travel narrative or toward a self-examination of time travel's emphasis on visuality. My rationale, for now—to be substantiated in the discussions that follow—is that such ahistoricism is itself historically warranted, given the diverse state of time travel narrative as it advances into the twenty-first century. My "Theoretical Interval," between Chapters 4 and 5, revisits the general outline of this topic once I have more clearly set up its theoretical parameters.

However, for the moment I wish to dwell on a matter I sidestepped earlier in the book for specific historical and conceptual reasons, that of H. G. Wells's *The Time Machine*. To recall, going against most literary-historical accounts, I declined to affirm Wells's story as a proper origin of time travel fiction. However, the phases of the history of time travel I have just outlined should help make clearer the role that Wells's book does play for later time travel writers and readers, and for the sorts of theoretical problems they confront. In brief, I argue that although the influence of *The Time Machine* may be considerable, it is also greatly belated, and awaits several cycles in the generic development of time travel fiction before it begins to hold sway.

The Time Machine

Wells's novella begins with a framing story narrated by an anonymous dinner guest visiting the house of a scientist referred to only as "the Time Traveller," who in turn relates his own story of a trip into the far future.[6] Arriving, approximately 800,000 years from his present, in a picturesque landscape littered with inscrutable ruins—yet identifiable as former London by the Thames River (among other landmarks), which "had shifted, perhaps, a mile from its present position"[7]—the Time Traveller encounters two types of beings, the subterranean Morlocks and the gentle, dissolute Eloi. Speculation about how this pairing of future humans might have arisen now occupies a good portion of the text, even though there are moments in which Wells indicates that the Time Traveller's theories may be considerably less cogent than his initially sanguine narration makes it appear. The Eloi and Morlocks have evolved "into two distinct animals"[8] from their earlier, more arbitrary segregation by social or economic class: "So, in the end," as the Time Traveller speculates, "above ground you must have the Haves, pursuing pleasure and comfort and beauty, and below ground the Have-nots, the Workers getting continually adapted to the conditions of their labour."[9] Elsewhere, the Time Traveller explicitly links this eventual disparity to "the merely temporary and social difference between the

Capitalist and the Labourer,"[10] and asks rhetorically, "Even now, does not an East-end worker live in such artificial conditions as practically to be cut off from the natural surface of the earth?"[11] Still, the Time Traveller somewhat spoils his dispassionate scientific appraisal with a nineteenth-century English aristocrat's spontaneous distaste for the subterranean Morlocks (the former "East-end worker[s]"), declaring that "[i]nstinctively I loathed them";[12] by contrast, he expresses a disquieting paternalistic attraction to the opposite "daylight race" of Eloi.[13]

After rescuing his time machine from the Morlocks, who are innately curious tinkerers, the Time Traveller continues on to the much further future, all the way to 30 million years hence, and ultimately witnesses the incipient heat death of the sun. These famous tableaux of cosmological decadence, which contribute to what Fredric Jameson calls "a grand narrative of entropy and dissolution," are precursors to the grandiose imagery of much twentieth-century science fiction:[14]

> From the edge of the sea came a ripple and whisper. Beyond these lifeless sounds the world was silent. Silent? It would be hard to convey the stillness of it. All the sounds of man, the bleating of sheep, the cries of birds, the hum of insects, the stir that makes the background of our lives—all that was over.[15]

Nonetheless, even such passages bear a close resemblance to descriptions of social evolution offered by earlier expositors such as Bellamy's Dr. Leete or Thomas's Professor Prosper. Despite the decisively alien qualities of the scenes, it remains vital to the theoretical stakes of Wells's plot that even this utterly forlorn future be a direct outgrowth of present-day development, in precisely the Darwinian sense familiar to the late-nineteenth-century utopian romance: a "rigorous projection of evolutionary biology," or a "putting [of] evolutionary theory into fictional practice."[16] Indeed, the Time Traveller continually withdraws his attention from the frenetic unfolding of his adventure to revert to a macrological exposition that attempts to explain—in a mode closely resembling the longwinded expositions of utopian hosts—both the physical and the social developments he is witnessing. As John Rieder notes, "*The Time Machine* explains itself at almost every step, from the opening lecture about time as the fourth dimension, through the Traveller's successive, shifting hypotheses about the origins of the Eloi and the Morlocks, to the astronomical theories that help explain the grim setting of the Traveller's final, bleak glimpse into the last days of life on earth."[17] Only Wells's insistence on continuous first-person narration rather than dialogue, along with the customary elegance of his prose, makes *The Time Machine* appear less manifestly wrought in its synthesis of apologue and macrologue than most other utopian literature of the period. In terms purely of the sheer quantity of explanation, its longwindedness, it is very much a typical late-nineteenth-century utopian romance.

Yet nearly all of the Time Traveller's attempts at evolutionary theory are halting, dubiously objective, and almost comically rife with self-doubt: "Very simple was my explanation, and plausible enough—as most wrong theories are!"; "My explanation may be absolutely wrong"; "It may be as wrong an explanation as mortal wit could invent."[18] Indeed, the Time Traveller alerts the reader—presumably alongside Wells's own ironic wink at the notorious unrealism of utopian apologues generally—that any proffered accounts of future evolution cannot be verified because "I had no convenient cicerone in the pattern of the Utopian books."[19] In this respect, *The Time Machine* is an attack on at least some core conventions of other contemporary romance, a "sarcastic intervention into th[e] utopianism" of writers like Bellamy and William Morris.[20] Even the longstanding critical tendency to identify Wells's book as a novella rather than as a romance (either "negative-utopian" or "scientific") is likely due in part to Wells's explicit disdain for the typical, clunky dialogic structuring of utopian romance, a disdain also evident, albeit in a different mode, in the byzantine expositions of *A Modern Utopia*.

In fact, in the version of *The Time Machine* that Wells published in the *National Observer* one year earlier, the plot unfolds far more in line with the "Socratic" structuring of typical utopian romance, as a dialogue between the already-returned Time Traveller and the more or less skeptical dinner guests. In the final 1895 version, Wells obviates that more awkward format by having his Time Traveller insist that the guests "refrain from interruptions" until the story is completed.[21] Nonetheless, in terms of its macrological apparatus, *The Time Machine* remains entirely continuous with the modes of uneasy realism that characterize the framing of late utopian romance, and for essentially the same reason: to furnish a realistic mechanism for the temporal transportation required of an evolutionary narrative, and to insert the viewpoint of an embodied and believable observer into a directly extrapolated future.[22]

The time machine itself, of course, is Wells's narratological coup, a superb means of interjecting the verisimilitude of mechanical engineering into an otherwise perilously fantastic narrative. Wells makes the machine move just as simply through time as a horseless carriage or steam car would move through space, merely by the switching of levers, the adjusting of verniers, and the applying of brakes,[23] and a large part of the macrological apparatus of *The Time Machine* is devoted to expounding upon the machine's mechanical exploitation of "the Fourth Dimension."[24]

However, the real innovation of the machine is its capacity to provide a concrete, *visual* confirmation of the evolutionary continuity of all history, a confirmation that arrives in the guise of a quasi-cinematic special effect:

> I pressed the lever over to its extreme position. The night came like the turning out of a lamp, and in another moment came to-morrow. The laboratory grew faint and hazy, then fainter and ever fainter. To-morrow night came black, then day again, night again, day again, faster and faster still. . . .
> . . . The dim suggestion of the laboratory seemed presently to fall away from me, and I saw the sun hopping swiftly across the sky, leaping it every minute, and every minute marking a day.[25]

Here, the direct observation of the external world provides an empirical demonstration of what the Time Traveller claims in theory, that "Time is only a kind of space"[26] and that therefore, given the right technology, a person could "stop or accelerate his drift along the Time-Dimension, or even turn about and travel the other way."[27] Indeed, the mere press of a lever accomplishes the task of advancing, reversing, or modulating temporal movement, as it would, for instance, in a kinetoscope, a protocinematic device with which Wells was very familiar.[28]

Visual proof is of paramount importance to the Time Traveller in Wells's final version of the story. He is so eager to demonstrate, for his somewhat indifferent dinner guests, an "experimental verification"[29] of fourth-dimensional travel that he sacrifices a scale model that "took two years to make"[30] simply in order to have them witness the miniature machine irretrievably vanishing "into the future or the past—I don't, for certain, know which."[31] This episode with the scale model—added only to the final 1895 narrative—also serves the ancillary purpose of inviting us to think about the precise appearance of the machine itself, since Wells is uncharacteristically thrifty in describing it. "[T]his lever," the host explains, "being pressed over, sends the machine gliding into the future, and this other reverses the motion. This saddle represents the seat of a time traveller."[32] Elsewhere we get glimpses of a "glittering metallic framework" with "ivory in it, and some transparent crystalline substance," and much later, a brief peek at "a thing of brass, ebony, ivory, and translucent glimmering quartz," possibly resting on sledlike "rail[s]."[33] At times, some part of the machine exhibits an "instability" or an "odd twinkling appearance . . . as though it was in some way unreal."[34] The interior contains a panel of clocklike "dials," one of which "records days, and another thousands of days, another millions of days, and another thousands of millions."[35] And that is virtually all; we barely know its overall shape.

Yet the most salient feature of the time machine, for the visualization of temporality that Wells endeavors to narrate, is already implicit in the initial description of the scale model: the simple fact that it is an *open* contraption. We realize this because the Time Traveller, or one of his guests, is able to reach inside the model in order to flip the tiny lever that sets it in motion.[36] And if one can reach into the machine, then presumably the rider who sits in the "saddle" is also free to look out, a feature crucial for the verisimilitude of Wells's story. Subsequently, nearly all the macrological energies of *The Time Machine* ignore the machine itself and devote themselves instead to describing what the Traveller himself sees from his seat: the essentially kinetoscopic "velocity through time"[37] of the accelerated objective world. This visual tableau starts, somewhat comically, with a view of the Time Traveller's housekeeper:

> Mrs. Watchett came in, and walked, apparently without seeing me, towards the garden door. I suppose it took her a minute or so to traverse the place, but to me she seemed to shoot across the room like a rocket.[38]

Mrs. Watchett's accelerated motion is like an exaggeration of the movements of a 12- or 16-frames-per-second silent film projection, and overall the scene is remarkably "cinematographic," as Wells will describe his later *A Modern Utopia*—or, more properly, kinetoscopic, in deference to the moving-image technology most widely available in the mid-1890s. What the Time Traveller sees is something like what the later viewer of a more high-tech time travel film might see: the compression, dilation, reversal, and potential doubling of narrative lines, produced in the form of a special effect but nonetheless narrated as fully part of the realist *fabula*. Upon returning from his voyage, the Time Traveller observes Mrs. Watchett walking across the room in reverse, precisely like a kinetoscope run backward: "[N]ow her every motion appeared to be the exact inversion of her previous ones. The door at the lower end opened, and she glided quietly up the laboratory, back foremost, and disappeared behind the door by which she had previously entered."[39]

This emphasis on moving imagery, as well as on its potential manipulation and reordering, is conspicuously absent from the earlier versions of *The Time Machine*, for instance the 1888 "The Chronic Argonauts," despite the fact that the fourth dimension and the unity of space and time are already theorized. What the new notion of an open machine enables Wells to depict is not so much the specific spatialization of time or history as its specific visualization, which now facilitates a stock of special effects invoked to create, for the Traveller, a sort of epic movie of future time:

> I saw trees growing and changing like puffs of vapour, now brown, now green: they grew, spread, shivered, and passed away. I saw huge buildings rise up faint and fair, and pass like dreams. The whole surface of the earth seemed changed—melting and flowing under my eyes.[40]

For a long while, the Time Traveller merely stays put, silent and invisible in his saddle, watching the succession of moving images that flows before him, as though sitting in front of the kinetoscope window. Garrett Stewart suggests that Wells's contraption is "as accurate a characterization of the machine of cinema as we could want." [41] And indeed, with extraordinary prescience, the machine looks like a kinetoscope turned inside out. One sits *inside* the device, presses the lever, and looks out through the "viewer"—an uncanny anticipation of the odd inversion of Edison's Black Maria filming studio that constitutes an early movie theater, in which the film image is reproduced on a screen by virtue of the projector's reversal of the original direction of light through the lens. In these parts of the story, it hardly matters whether it is the Time Traveller himself who moves through time, or time that moves across the screenlike "display" of his visual field:

> What strange developments of humanity, what wonderful advances upon our rudimentary civilization, I thought, might not appear when I came to look nearly into the dim elusive world that raced and fluctuated before my eyes! I

saw great and splendid architecture rising about me, more massive than any buildings of our own time, and yet, as it seemed, built of glimmer and mist. I saw a richer green flow up the hill-side, and remain there without any wintry intermission.[42]

Eventually, as the machine's "velocity" surpasses the scale of any human history, the cosmos itself becomes a cinematographic show, and the narrative starts to look like an experimental time-lapse film. Astrophysical movements are projected as a signifying interplay of light and dark:

As I drove on, a peculiar change crept over the appearance of things. The palpitating greyness grew darker; then—though I was still travelling with prodigious velocity—the blinking succession of day and night, which was usually indicative of a slower pace, returned, and grew more and more marked. . . . The circling of the stars, growing slower and slower, had given place to creeping points of light. At last, some time before I stopped, the sun, red and very large, halted motionless upon the horizon.[43]

In essence, the time machine has become a prototype of cinematic infrastructure, still in service of an "empirical verification" of evolutionary history, but now far surpassing that original goal in its ability to depict time itself, and the literal perspectives from which we might view it. And it is only in this specific aspect—as the early, literal depiction of a machinery of narratological viewpoint—that *The Time Machine* is a vital precursor to later time travel fiction in its "multiverse/filmic" phase, although one would have to jump several decades forward to see its full effects.[44]

Finally, then, one can explain what might otherwise, from the retrospect of more contemporary time travel fiction, seem like an anomaly in a writer otherwise so deeply interested in redepicting and manipulating temporality: Wells's apparent disinterest in possible causal contradictions or paradoxes. This disinterest is all the more striking if one notices that Wells seems perfectly cognizant of their possibility. The Time Traveller's dinner guests momentarily offer conjectures about potentially paradoxical scenarios, such as a historian who "attract[s] attention" while attempting "to verify the accepted account of the battle of Hastings," or the possibility that one might "invest all one's money, leave it to accumulate at interest, and hurry on ahead!"; the narrator himself mentions his passing interest in "the curious possibilities of anachronism and of utter confusion [time traveling] suggested."[45] A number of characters even refer to the Time Traveller's experiment bluntly as a "paradox," although it is clear they do not mean the sort of acausal loop that is the special province of the later time travel story; rather, they refer only to a straightforward conceptual or scientific impossibility, or at least to a possibility not yet rationally explained.

In short, Wells is entirely unconcerned—or it doesn't occur to him to be concerned—with the machine's potential to cause the sorts of temporal alterations that will eventually make time travel into paradox fiction, and that just begin to be hinted at by later writers such as Mackaye and Wertenbaker. But indeed, why would Wells be interested in paradox, when the machine's sole function is to depict an *actual* image of time in its evolutionary unfolding? Only the Time Traveller's dinner guests, as yet unacquainted with the technology required to produce such an image, are sufficiently uninformed to suggest that time travel might entail paradox. The Time Traveller himself, like Wells, is concerned only with a fully self-consistent exhibition of temporal movement, and with its use in constructing a credible model of both social and cosmological history—the same tacit theoretical goal of all Bellamy-era utopian romance.

In this sense, Wells's *The Time Machine* fits decisively within what I have called the first phase of time travel fiction, in which time travel devices belong to a macrologue, subordinate to the aesthetic and theoretical goals of utopianism. Notwithstanding the importance of *The Time Machine*, both for the development of *fin-de-siècle* scientific romance into science fiction, and as a precedent for much later spacetime machinery, the book has no direct influence upon time travel fiction per se until long after the subgenre has solidified into its recognizable position within the larger realm of popular science fiction. In this sense, preceding the post-utopian epoch of Mackaye—and I am well aware how counterintuitive this claim will sound from the perspective of a more conventional genre history of either time travel or science fiction generally—*The Time Machine* is *not* yet a time travel story; rather, it is an especially well-written and cleverly theorized utopian romance.

THREE

"The Big Time": Multiple Worlds, Narrative Viewpoint, and Superspace

Early in Albert Einstein's famous 1905 paper "On the Electrodynamics of Moving Bodies," the first formulation of his special theory of relativity, he remarks that any assertions of electrodynamic theory must "concern the relations between rigid bodies (coordinate systems), clocks, and electrodynamic processes."[1] "Insufficient regard for this circumstance," Einstein then observes, "is the root of the difficulties with which [the] electrodynamics of moving bodies must currently struggle."[2] By insisting on observations and measurements only of "rigid bodies and clocks" rather than of metrical abstractions such as centimeters and seconds—or, in other words, in Max Born's paraphrase, by insisting on the positivist "heuristic" rule that "concepts and statements which are not empirically verifiable should have no place in a physical theory"[3]—Einstein is able to articulate and eventually resolve, in extremely elemental terms, several crucial problems that had bedeviled electrodynamic theory since Maxwell. In essence, Einstein's physical rather than geometrical interpretation of measurement strictly respects the fact that the velocity at which information propagates is limited by *c* (the velocity of light), and that therefore the relative velocities and remoteness of observers will influence data these observers gather about events at any non-negligible distance in real space and time. The upshot is that observers in different reference frames, or "coordinate systems," will ineluctably disagree about where and/or when events have occurred. The metaphysical concept of simultaneity must therefore be altogether discarded, an eventuality that Born, among many others, describes as "a new formulation of the fundamental properties of space and time."[4]

In his 1916 book *Relativity: The Special and General Theories*, Einstein elaborates more fully, now in prose rather than mathematics, what his emphasis on "rigid bodies and clocks" entails for our understanding of the relation between physics and geometry. Again beginning not with relativity per se, indeed not with physics at all, Einstein asks us, in the mode of a grade school mathematics teacher, to consider what we mean when we say that geometrical propositions are "true." Within any geometrical system, we assert the truth of particular propositions—for instance, in the Pythagorean theorem, that the square of the hypotenuse of a right triangle equals the sum of the squares of the two sides—when we can deduce such propositions from the system's axioms by way of a proof. However, as Einstein asserts, when we then inquire about the reliability of the axioms themselves—for example, when we ask whether it is the case that only one straight line, such as the line we have just used to construct the right triangle, goes through any two points—not only can there be no true-or-false answer, but the question "is in itself entirely without meaning."[5]

The phrase "entirely without meaning" is somewhat excessively rigorous in its positivism. Einstein acknowledges his own hyperbole, suggesting that despite what we may know in theory, we still "feel constrained to call the propositions of geometry 'true,'" which means we still think of them as "correspond[ing] to more or less exact objects in nature."[6] Here Einstein's reader encounters physics for the first time, in a rudimentary form appropriate to what Einstein calls our "habit of thought."[7] That is to say, the reader encounters a description of a physical world that seems to correspond, more or less exactly, to familiar Euclidean geometry: in a word, "nature." Still playing upon our ingrained "habit[s] of thought," Einstein then invites us to reconsider our geometrical axioms about points and straight lines strictly physically, as statements about actual material objects. Instead of an abstract line segment connecting two points, we will speak only of a "practically rigid body," a rod or stick, the two ends of which no longer merely correspond to, but actually *are* at a certain distance from one another within the physical world.[8] That distance is *ipso facto* the length of the rod or stick itself, which now becomes our default measuring unit for any other distance. As Einstein then states, "Geometry which has been supplemented in this way [can be] treated as a branch of physics."[9] And this puts us in a position, finally, to inquire legitimately as to the *truth* of our question about two points and a line, since we may now refer to real arrangements of locations on a "rigid body" that can be touched, manipulated, and physically observed.

In essence, Einstein has called a metaphysical bluff, one endemic to theoretical speculations of all sorts. If one is willing to talk about the geometry of spacetime, then one ought *really* to do it, not signifying abstract lines, points, and durations, but instead referencing only material bodies and actual locations upon those bodies. Born describes this as a "principle which certainly had been known before but which was used mainly for logical criticism and not for scientific construction . . . [,] a principle demanding the elimination of the unobservable."[10] Einstein invokes the principle, for instance, to critique our conventional use of the concept of time, inviting us to consider what we actually mean, empirically and materially, when we speak of the moment at which an event occurs. In purely physical terms, it turns out that what we mean is "the reading (position of the hands) of that one of these clocks which is in the immediate vicinity (in space) of the event."[11] Thus, as Einstein says, "the definition of time in physics" will now also be construed entirely physically: as the current position of yet another rigid body (a clock hand) attached to a physical device (a clock), discoverable in a spatial location negligibly close to some physical event. Einstein is here proposing a kind of elemental methodological rigor, presumably valid and necessary for any adequate inquiry into the geometrical rules that govern the properties and relations of objects: if one makes assertions about the physical world, one should be consistent and give these assertions themselves physical interpretations—we will speak not of "space" per se, but only of "practically rigid bodies"; not of "time" per se, but only of "mechanical clocks" and "hands" in physical locations.[12] Within physics (and its attendant philosophies), the surprising result of this simple bluff-call is the elimination of any conception of space and time independent of the actual physical masses that occupy them—in brief, the elimination of any space or time merely through which objects move, or merely in which they rest.

It is not Einstein's conclusions but rather the elemental methodological rigor alone—in a word, his adherence to a severe terminological *materialism*—that I wish to import from physics into literary theory, for the specific purpose of invoking it as a protocol for careful narratological reading. Pursuant to such a terminological materialism, the critic aspires to a self-discipline comparable to Einstein's: when I make statements about the conditions under which a narrative event takes place, I will try to give such statements strictly physical interpretations, refusing to abide by merely conceptual or metaphorical descriptions of what happens when a person reads a text or, as I shall later discuss, views a film. And I ought to be motivated by very much the same logical compulsion that drives Einstein in his own discussion. As a critic of literature or media, one inevitably deals with entities in an actual universe: not merely stories, fictional landscapes, and viewpoints, but instead readers, texts, eyes, and material arrangements of bodies.

As with any number of narratological inquiries, time travel stories will be especially helpful in such a methodological retooling because of the ways in which they portray, in literal fashion, the basic conditions of narrative experience; again, they are our narratological laboratory. In essence, time travel stories are the physical redescription, and therefore potentially the testing or retesting, of narratological axioms, a fundamentally materialist literature or, more precisely, a material-narratological literary type. In this precisely restricted sense, time travel is "Einsteinian" fiction—not because it is about physics or relativity, but rather because it strives to be rigorously physical in its basic narratological construction, and because it does so virtually by default.

Let me recommence my inquiry into time travel fiction in 1945, an important year for physics as well as for literature and criticism.[13] From there, I will trace some peculiar narrative problems within time travel fiction up to the present or close to it, and view them both in light of more current narrative theory and alongside more recent developments in theoretical physics.

A Physics of Narrative

Joseph Frank, in his well-known 1945 essay "Spatial Form in Modern Literature," analyzes an episode in Gustave Flaubert's *Madame Bovary*, a rendezvous between Emma Bovary and her lover Rodolphe at an agricultural fair, during which several distinct levels of narrative action occur simultaneously. By carefully juxtaposing these levels in terms of their relative proximities, their rhetorical tones, and ultimately what Frank calls their "spiritual significance," Flaubert is able to achieve a "devastating irony" when the novel's reader is finally able to hear, as the author himself describes it in a subsequent letter, "the bellowing of the cattle, the whisperings of the lovers and the rhetoric of the officials all at the same time":[14]

> "—Did I know I would accompany you?"
> "Seventy Francs!"
> "—A hundred times I tried to leave; yet I followed you and stayed . . ."
> "For manures!"
> "—As I would stay to-night, to-morrow, all other days, all my life!"
> "To Monsieur Caron of Argueil, a gold medal!"
> "—For I have never enjoyed anyone's company so much."
> "To Monsieur Bain of Givry-Saint-Martin."
> "—And I will never forget you."
> "For a merino ram . . ."[15]

Presumably any competent reader of Flaubert would be able to grasp the irony of this juxtaposition of voices. For the critic, it is rather more difficult to describe, in really precise terms, just how Flaubert manages to convey such irony within the confines of linear prose. Frank describes the technical difficulties that confront the novelist: "[S]ince language proceeds in time, it is impossible to approach this simultaneity of perception except by breaking up temporal sequence," jumping back and forth between the various "levels" so as to communicate to the reader, through some technique other than simple diachronic progression, everything that is simultaneously happening.[16] Thus, although the officials, the animals, and Emma and Rodolphe all carry on in tidy linear fashion, the reader, by contrast, is obliged to hop back and forth in order to assemble the multiple substories into a single coherent experience. Of course, given Flaubert's narrative skill, as well as the reader's own familiarity with classical conventions of temporal disarticulation and reconstruction, the resulting effect is decidedly not discontinuity or disorientation, but rather a unified impression. In brief, the reader gets a *scene*, in which, in Flaubert's words, "everything should sound simultaneously"[17]—and the getting of the multiple simultaneous temporalities of that scene is precisely what is required for its irony to come across. Frank calls what occurs a "spatialization of form," in which, as he says, "relationships are juxtaposed independently of the progress of the narrative; and the full significance of the scene is given only by the reflexive relations among the units of meaning."[18] Like the fragmented sequences of modernist poetry, to which Frank's analysis will subsequently turn, the full scope or significance of Flaubert's agricultural fair scene is understood only after its several distinct minihistories are "apprehended" together, in a posterior reading act that Frank names, conveniently but not very informatively, "reflexion."

Here is precisely the sort of sneaky metaphysical term that tends to emerge in narrative theory, and which I want to recast in accordance with the methodological discipline I obtain from Einstein's physical reinterpretation of geometry. Would it be possible to give Frank's spatiotemporal phenomenon of "reflexion" a wholly physical description, as Einstein does with acts of measurement? Flaubert's scene is understood by the reader seemingly *after* it is read, and Frank perhaps hastily identifies this after as a specific "moment of time" or "instant of time."[19] The problems with such a description are immediate and confounding. One might start simply by noting the discrepancy between the rather technical, quasi-metaphorical Latinate terminology with which Frank names this "instant"—"reflexion," "spatialization," "apprehen[sion]"—and the relative rapidity, even effortlessness, of the reader's actual comprehension, not to mention the easy postulation of the "space" in which that comprehension occurs. I can then ask, in sympathy with our easily comprehending reader, some straightforward but also more precisely physical questions about the reader's experience in space and time, very much as Einstein does about the famous observer of lightning strikes on the train tracks:[20] Just how soon does the reflexive reconstruction of meaning, its juxtaposition of actions and vocabularies, occur? Just where, in the course of the story's unfolding—that is to say, at what point in the diachronic progression of the plot through the text—can reflexion be said to happen, or to have happened? Does the experience happen just once, like an epiphany—Flaubert himself rather casually states that it occurs "all at the same time"—or does it actually develop progressively over the course of the reader's literal movement across the lines on the page? Indeed, does reflexion have an "instant" at all, or only the retroactive illusion of one, even a retrospective projection, like the reconstituted primal event in psychoanalytic discourse? Despite the ostensible simplicity or unity of the reader's assimilation of the completed scene, no simple, physical answers to such questions about its construction or reconstruction are forthcoming for the critic. Or rather, ironically, the only available answers to such spatiotemporal questions about the actual reading of a narrative text would seem to be, so to speak, transcendental rather than phenomenal or empirical: reflexion will (at some indefinite point) already have occurred while the reading was being enacted, or the reflexion always will have underlain the (now completed) act of reading, or the reading act will already be (or will already have been) reflected upon, and so on. In short, Frank's supposed "act" of reflexion, necessary and palpable though it appears to be for the scene's properly ironic reception, nevertheless has no single or simple spacetime, certainly no "instant," but instead occurs always from a multiplied position, one with several different presents and pasts, empirically locatable neither in the text itself nor in the reader's life.[21] To borrow terms remote from any conventional literary-critical discourses about nineteenth-century fiction, but that may still turn out to be surprisingly appropriate, *Madame Bovary* can be read only in hyperspace and hypertime.

Since the time of Frank's essay, literary theorists have developed more nuanced terms for describing the structurally retroactive process of reflexion that occurs in reading comprehension. Theorists have long since come to take for granted that a narrative fundamentally "claims to repeat a journey already made"[22] or "presents itself as a repetition and rehearsal . . . of what has already happened,"[23] a pretense of reiteration that dislocates the reader's experience from any particular line of events occurring within the story. In narratological analysis, this suggests what Wolfgang Iser calls a "dialectic of pretension and retention," one that any reader must negotiate in order to understand a literary text. Iser notes that because "the 'object' of a text can only be imagined by way of different consecutive phases of reading," entailing multiple perspectives, the reader's complex position is something like a "wandering viewpoint" that "is not situated exclusively in any one of the perspectives . . . [but] can only be established through a combination of these perspectives."[24] Nevertheless, the problem that lies at the core of Frank's analysis—to locate, as it were physically, the spacetime in which the disparate times of a multiplied narrative are brought together and reconciled into a single experience—remains sufficiently pressing and difficult that, even in current narratological criticism, it is not yet clear how such a space might be elucidated. Is the space an artifact of narrative form, or a part of the reader's phenomenological response to that form, or an underlying structure of narratives generally, or even a structure of the subject itself, of the reflective "I," which is also an "it," and therefore able to (re-)view its own history of reading? The residual spatial metaphorics of Iser's own language—"wandering," "position," "situated," and of course "viewpoint" itself—shows us that a fully precise description of where or when the reader is located while reading the text remains elusive and bound to quasi-allegorical—or what Einstein might have called geometrical—figurations that do not get fully translated into their literal or physical counterparts.

And oddly enough, despite the theoretical complexity of this problem of locating the reader's position, it is quite obvious that we occupy such hyper- or metatemporal spaces all the time, even during quite mundane narrative situations—for instance, every time we tell or hear a story with the structure of "meanwhile . . ." or "at the same time. . . ." The difficulty of accounting theoretically for this mundane narratological position may well be due to the intimacy with which reflexion structures our most basic encounters with historical or narrative space and time, the sort of difficulty one has describing a thing too close, like the muscular movements of one's own eyes. We really live, for all practical purposes, in hyperspace and hypertime, and tell our stories from within it. I would not presume fully to solve any such problem outright, but rather only to contribute something to its ongoing contemplation and complication. In that spirit, I offer the minor observation that the extratemporal reflexion that Frank discovers, first in Flaubert's agricultural fair scene and then more extensively in the experimentations of Proust, Joyce, and Barnes—a reflexion that, to some degree, occurs every time a narrative contains more than one simultaneous line—looks very much like what a time traveler does.

How might this characterization aid in providing a physical description, or what a physicist might call a "classical" description (possibly congruent with Genette's references to classical narrative), of the reading act? Here is one way: with a little tweaking of the critical language, it would be easy to reimagine the structure, if not the content, of Flaubert's scene newly cast into the plot of a time travel story, with the reflexive reader as the time-traveling protagonist. One might then hypothesize a fully physical description of what that reader does in the process of comprehending a given scene: through some technological mechanism of intra-temporal comparison— say, a time machine—I now travel several times through the multiple spacetimes of the agricultural fair, all the while sitting in a physical apparatus, the hyperspatial or hypertemporal location of which permits me access to multiple simultaneous lines of events, the full range of which I am now in a position to compare or ironically juxtapose when I arrive back "home." What I discover in this new time travel adaptation of the reading of Flaubert is not a transcendental condition, reflexion, but rather a literal plot: several sequences of events overlapping or repeating as I, their literal viewer—Frank, in a Wellsian moment, describes Flaubert's scene as "cinematographic"—sit in the saddle of my machine, traveling successively back and forth through time until I've got it all.

The point of this slightly flippant rewriting of Flaubert is not, of course, to suggest that *Madame Bovary* ought to have been published in *Argosy* or *Amazing Stories*, but rather to propose that a consideration of certain aspects of popular time travel plots, particularly the metatemporal "spaces" literally occupied by their protagonists as they move through or across multiple lines or worlds, can shed some light on the complex yet common reflexion needed to read *Madame Bovary* or any story like it. The time travel version of Flaubert's scene is our laboratory plotting of the act of reading, in which the problems of reflexive space and of narratological experience generally are experimentally worked through, now in the guise of literal objects and actions, and by physically present characters, rather than by metaphysically unsituated narrator-readers.

We can continue to experiment: if reflexion is like time travel, let us locate some time travel stories that already describe the hyperspace and hypertime of narration, and see how they do it. I will start with a story that dates from roughly the same period as Frank's essay: Jack Williamson's novel *The Legion of Time*, published serially from May through July 1938 in the pulp magazine *Astounding Science-Fiction*.[25]

Narrating from Hyperspacetime

The plot of *The Legion of Time* can be summarized quickly without doing it much harm. The story's hero, a young engineering student named Denny Lanning, is visited by "two anachronistic women" who have traveled back in time from competing possible futures: Lethonee of Jonbar—blond, "doe eyed," and "clad all in white"—and Sorainya of Gyronchi, sporting crimson chain mail and red fingernail polish, a kind of futuristic Bettie Page. The virtuous Lethonee recruits Denny Lanning, along with a "legion" of other past soldiers, to fight for the existence of Jonbar, since, as Williamson observes, "either Jonbar or Gyronchi—either Lethonee or Sorainya—may exist. But not both."[26] In other words, as in many such stories, a rift in the spacetime continuum must be healed, and only one possible future preserved. From here, the plot about a war between the two futures turns rather thin, and Williamson is able to protract it to novel length only via the ridiculous premise that the evil Sorainya of Gyronchi is "too lovely to be slain" by the male protagonists, notwithstanding the effectiveness of their weapons against her army of mutant anthropoid ants.[27] In the end, Gyronchi is defeated and all the incompatibilities between the timelines are resolved. The conclusion of the novel has Denny Lanning, in an adolescent straight boy's apotheosis, acquire a literal composite of Sorainya and Lethonee "fused into the same reality." Williamson ends: "He drew them both against his racing heart, breathing softly, 'One!'"[28]

I wish to give some fuller attention to Williamson's description of the relationship between the story's two possible futures, a description couched in a terminology of spatiotemporal dimensions that combines physics and narratology far more effectively than Williamson combines character and action. In the following passage, Williamson cites terms from a scientific textbook that Denny Lanning has been reading, entitled *Probability and Determination*:

> The elementary particles of the old physics may be retained, in the new continuum of five dimensions. But any consideration of this hyper-space-time

continuum must take note of a conflicting infinitude of possible worlds, only one of which, at the intersection of their geodesics with the advancing plane of the present, can ever claim physical reality.[29]

Correlating the probabilistic branching of the novel's own plot with "the new quantum mechanics," Williamson goes on explicitly to reference Planck, de Broglie, Schrödinger, and Heisenberg. Thus the structure of the time travel story is generated, with at least a guise of scientific rigor, from current theoretical physics: it is precisely this five-dimensional continuum through which the protagonists will have to journey in their time-traveling ship the *Chronion* in order to visit the two futures they successively encounter.[30] In other words, the time ship does what narratives generally do, and what Flaubert's reader also does—travel through and across four-dimensional spacetimes, or juxtapose them in order to adjudicate their relationship—but it does so literally, in a physically mobile machine. Moreover, the medium through which this machine accesses the novel's two alternate future spaces is a third space, a kind of "superspace," or what Lanning's physics textbook calls a "hyper-space-time," which the novel now undertakes to describe:

> The ship must be moving. But where?
>   Looking about for a glimpse of the sun or any landmark, Lanning could see only a curiously flickering blue haze. He went to peer down over the rail. Still there was nothing. The *Chronion* hung in a featureless blue chasm.[31]

This metaworld or "hyper-space-time," a strangely ambiguous structural no-place, a "nothing" described as "featureless" and "queerly flat," is nonetheless put together just like any normal narrative spacetime, with a diachronic progression and more or less conventional spatial features: gravity, an unambiguous sense of up and down ("He went to peer down over the rail"), a breathable atmosphere, ambient light, and so on. Its status thus seems to partake of two very different characterizations. First, it is like a transcendental structuring condition, as it were a necessary syntactical connector or "super-" setting required to link the two depicted possible worlds of the novel. Second, it is like a straightforward milieu, a scenario no different in its basic spatiotemporal arrangement from the future cities of Gyronchi or Jonbar themselves, or from the "hushed April evening of 1927" in Cambridge, Massachusetts, when Denny Lanning first encounters the time-traveling Lethonee of Jonbar.

Here, then, we have our physical description of the narrative superspace of reflexion, and of the comprehension of juxtaposed narrative lines that it permits. It is the equivalent of Einstein's insertion, into the discussion about the measurement of spacetime events, of actual rigid rods and mechanical clocks: the physical literalization of the *geometry* of storytelling and of its axiomatic conditions. In narratological terms, any pair of possible fictional worlds, if a protagonist is going to travel between them and comprehend their relationship, demands such a superspace, a *hyperspacetime* to link them through a further narrative discourse and movement. Without it, to describe things physically rather than metaphysically, the protagonist's perspective or viewpoint could only shift instantaneously from the first possible world to the second, perhaps in the mode of a hallucination or formal trick on the page, but not in such a way that either the character or the reader could undertake a reflexion to consider them as interrelated components of a single larger story. In this sense, the hero's situation, journeying through physical hyperspacetime in the time-traveling ship *Chronion*, is no different, aside from its literalness, from the situation of any reader or viewer of a fiction—for instance, the reader of Flaubert who "travels" to (or across) possible fictional worlds, all the while implicitly negotiating the relationship between such fiction and the real milieu of the reading act itself.

Science fiction authors have depicted quasi-narratological superspaces in a variety of ways, a few of which I will briefly describe. In Fritz Leiber's 1958 *The Big Time*, from which I have borrowed my chapter's title, Leiber tells the story of an immensely prolonged "Change War" fought in some indeterminate period of the future by mysterious time-traveling organizations known only as the Spiders and the Snakes.[32] Combat in a Change War consists of attempts to gain strategic advantage by traveling backward and forward through time and manipulating pivotal historical events, for instance, "poisoning Churchill or Cleopatra" or "kidnapping Einstein when he's a baby."[33] The final outcome of the war will not be known until "a billion or more years from now," when one "victorious" historical line will have successfully encompassed whatever events or consequences are sufficient to eliminate all possible alternatives.[34]

The narrator of Leiber's story is Greta Forzane, an "entertainer" (which is to say a prostitute) for Change War soldiers; she works in what she euphemistically calls a "Way Station." Here is where things begin to get narratologically interesting, however sociologically or generically provocative they have already become. The Way Station is located in an artificially created nonspace between embattled possible worlds, an interstice Greta calls "the Place":

> [M]y job is to nurse back to health and kid back to sanity Soldiers badly roughed up in the biggest war going. This war is the Change War, a war of time travelers—in fact, our private name for being in this war is being on the Big Time. . . .
>
> The place outside the cosmos where I and my pals do our nursing job I simply call the Place. A lot of my nursing consists of amusing and humanizing Soldiers fresh back from raids into time.[35]

"The Place," like Williamson's "featureless blue chasm," is strangely ambiguous. Disconnected from any particular chain of events or worldlines in the cosmos, "the Place" is a metahistorical or possibly extrahistorical location, strictly speaking not part of the fictional world that the novel posits as its historical setting, although of course it remains fully part of the story's *mise-en-scène*. In this aspect, "the Place" somewhat resembles, in its position with respect to the rest of the plot, what Genette calls a "paratext," a subsidiary or supporting space of narration "outside the cosmos" of what the novel constructs as actual history.[36]

Yet despite its existential ambiguity, "the Place" ends up looking very much like any other narrative scenery, and adheres both to a conventional ordering of time and space and to the conventional generic patterning of Leiber's novel overall. Greta even describes "the Place" as a series of rooms "midway in size between a large nightclub . . . and a Zeppelin hangar," rooms that, aside from their peculiar ontological status—that is, aside from the extraordinary aspect of their location "on the Big Time" rather than in the "Little Time[s]" of particular worldlines—are not much different from the rooms one might find in any typical intracosmic brothel.[37] Indeed, if Greta were simply to retranslate her science-fictional jargon back into more conventional language, as I briefly did with a scene from Robert Silverberg's *Up the Line* in the Introduction—"cosmos" back into "city street," for instance, and "the Void" (the barrier between "the Place" and the cosmos) back into "wall" or "doorway"—the story could well be mistaken for the kind of *noir* war satire with which it shares much of its action and tone. The crew of battle-weary soldiers who, in the first scene, burst through the "Void" into "the Place" could be characters from a knockoff of *What Price Glory?* or *Catch-22*, bursting through the front door of their favorite bar or brothel at the start of a bender. By comparison, and closer to generic home, a character in John Crowley's *Great Work of Time*, having "exited from time and entered a precinct outside it," nonetheless finds himself, in this "not quite existent" space, encountering "the solidity of its parquet floor and the truthful bite of its whiskey."[38]

Time travel writers often endeavor to describe extranarrative, quasi-existent locations similar to Leiber's "Place" or Williamson's "chasm," locations wholly outside of space and time or, more precisely, outside of any specific sequence of historical events. Clifford D. Simak depicts a "Highway of Eternity," a "never-ending thread of road" situated in an utterly empty, gray landscape, "yanked askew and canted out of normal time and space"; in another novel, Simak writes of a "foggy nothingness . . . striving to camouflage its very fact of being."[39] Greg Bear has an infinitely long "Way"; Roger Zelazny "the Road [that] traverses time"; Michael Moorcock a "subspace" or a "megaflow"; Robert Silverberg an "ultraspace," a "gray, featureless murk that doesn't change at all, ever"; and Ray Cummings "a formless vista of Nothingness."[40] John Brunner describes a "curious skew-axis of hypothetical or speculative time—the medium in which existed improbable alternative worlds," and interestingly suggests that any alteration of history occurring within this medium would constitute what he calls a "sluggish event."[41] Metaphors of tunnels and liquids abound, as these are convenient figurations of linearity, unidirectionality, or thermodynamic "flow": Greg Bear refers to "chronological backwash"; John Brunner, "a welling down the ages . . . like an incoming tide" or an "onrushing tide of consequences" or even "an avalanche."[42] L. Sprague de Camp characterizes temporal "influence as a set of ripples spreading over a pool," while Peter Heath depicts a familiarly Heraclitean "infinite river of time," as the physicist Igor Novikov will later do; F. M. Busby refers just to "river-time."[43]

Brian Ball's *Timepivot* combines several metaphorical conventions to create a "Forever Planet" on which "there was no sense of *forward*, or *down*, or *through*, though the sinking succession of tiny universes flowered ahead, below and tunnel-like."[44] This intriguing depiction evokes with some detail the incongruity between the extraordinary ontological status of any such "Big Time" location and the rather ordinary narrative sequencing, itself undeniably both spatial and temporal, of experiences occurring within it:

> [The Forever Planet] was the interval between two times, the space empty of space. . . . There was no time sequence, so events could not progress in a temporal linear fashion. *Yet there was a trend in events!* . . . *There was still personal identity.*[45]

What is incongruous about such nonspaces or nontimes within time travel literature is precisely their congruity, the manner in which they duplicate conventional spatiotemporal conditions. Despite their apparently radical removal from universes of history, human psychology, and storytelling, they nonetheless remain entirely suited to regular science fiction narrative, structured and formed according to the usual generic rules. As narrated, the nature of "the Big Time," superspace, or hyperspacetime is itself necessarily spatiotemporal, and therefore becomes a literal setting of the story. This literalness, I am suggesting, is the specific theoretical advantage that the popular time travel writer holds over the spatiotemporal machinations of literary experimentalism, for instance the self-consciously ironic and parabolic descriptions of Borges's "The Garden of Forking Paths" or the spatiotemporally ambiguous narrative sequences interspersed and italicized between ongoing scenes in Virginia Woolf's *The Waves*.

A last science-fictional example, for the moment: Isaac Asimov's "The End of Eternity."[46] "Eternity," as Asimov depicts it, is a kind of elevator shaft traversed by temporal agents ("Eternals") in a special vehicle called a "kettle":

> The kettle didn't pass through space in the usual sense of the word, and of course it didn't pass through Time, since Eternity short-circuited all of Time from the 28th Century . . . to the un-plumbable entropy-death ahead.
> But Father Time! the kettle went through or over or along *something*.[47]

Despite the kettle's not passing through either space or time, the tubelike region through which it does run is configured in sufficiently linear fashion as to make all of the Eternals' descriptions of the kettle's one-dimensional movement far more literal than metaphorical—in fact, they are analogues of the historical timeline one might see on a schoolroom chart. As Asimov says in another story, "passage through time itself takes time."[48] Thus, for instance, the kettle "pass[es] up the corridors of the endless centuries," "fit[s] snugly inside the vertical shaft," moves specifically "upwhen" and "downwhen," and "arrives" and "departs."[49] Eternity, for all practical (that is, for all narrative) purposes, has its own fully fledged spacetime, which operates as a more or less explicit parallel to "real" historical spacetime, progressing alongside the "line" of normal centuries and separated from actual history by a threshold described in terms that borrow simultaneously from quantum physics and from what Asimov would have perceived as a quasi-mystical metaphysics:

> The barrier that separated Eternity from Time was dark with the darkness of primeval chaos, and its velvety non-light was characteristically speckled with the flitting points of light that mirrored sub-microscopic imperfections of the fabric that could not be eradicated while the Uncertainty Principle existed.[50]

The ambiguity between physics and metaphysics in Asimov's description of this "barrier" is just right. In the jargon of "hard" science fiction, such a language almost always signifies the limit of a writer's (or a genre's) current explanatory capacity, wherefore metaphysical language is summoned in the face of a future invention or discovery that, for the present-day reader, cannot yet attain its fitting terminology.[51] Nothing in this metaphysical language is therefore inconsistent with the realist ambitions of the "hard" science fiction writer, who uses such language precisely at the point where he or she must, realistically, acknowledge the extreme threshold of a scientific vocabulary, the moment at which, as Arthur C. Clarke's often cited "third law" states it, "any sufficiently advanced technology is indistinguishable from magic."[52]

But the difficult description of a barrier between Time and Eternity is in fact a canonical metaphysical problem, and although it is unlikely that Asimov is fully aware of his own Neoplatonism here, his language echoes both that of Plotinus—"We must then have, ourselves, some part or share in Eternity. Still, how is this possible to us who exist in Time"—and of Augustine—"O Lord, since you are outside Time in Eternity, are you unaware of the things I tell you?"[53] These comparisons are not intended to elevate Asimov's speculative efforts beyond their due, but instead to indicate that the terminological limit he encounters is no mere coincidence of the futuristic setting of the science fiction novel, nor merely an artifact of the generic constraints of a particular mode of storytelling. Rather, it is the result of a more fundamental failure of description that the time travel novel exposes at the juncture of several basic narratological or philosophical problems: the nature of historical time, the nature of psychological perspective, and the essential contribution of both these first two problems to what we may now begin to perceive as the overarching problem of the self-depiction of narrative form. At the barrier between time and its other, the narratological structure of time itself emerges, here rendered literal rather than allegorical or structural.

So here again, in Asimov, as in all the writers I have mentioned who set out to depict literally the hyperspacetime between spacetimes, we have what Frank calls the "spatialization of form," however not as some experimental or theoretical literary device but rather as sheer plot: the superspace of the act of reading or of observing narrative events, historical lines, or fictional worlds (or parts of worlds), described as a literal "place," "corridor," "tunnel," "road," "chasm," "river," or "shaft"—a location from which, as a Robert Heinlein narrator remarks, we "stand off and perceive duration from eternity."[54] Time travel fiction is in this respect a mimetic rendering of "the Big Time," a realistic portrayal of that superspatial perspective required of any narrative juxtaposition and, in turn, of any philosophical or quasi-philosophical inquiry into the structure of worldlines, beyond bare mathematical formalism. In such a light, we ought to consider fully resurrecting that term often swiftly dismissed by narrative theory for its seeming vagueness and its contamination by problematic ocular metaphors: *viewpoint*. What time travel stories demonstrate is: (1) the transcendental necessity of a superspace in any narrative rendering of time; (2) the fundamentally *visual* character of that superspace, underscored by the descriptive normality of its spacetime and the sense in which we merely occupy it like any other narrative scenery; and (3) the precedence of popular, rather than experimental, narrative in positing this quasi-transcendental spacetime as a *real* milieu of the reader, the interworld of his or her perspective. The time travel story literally depicts the physical conditions of "the Place" where the "points" from which we "view" plots unfolding must be presumed to abide.[55] In the chapters that follow, I will have considerably more to say about how viewpoints operate, and why their visual metaphorics ought not to be too readily disparaged, even with respect to purely textual media.

Superspace

Because this mode of theorizing narrative superspace or viewpoint—for that is what I am claiming popular time travel writers do—is literal rather than phenomenological or transcendental, it can be carried to quite fascinating extremes. For example, late in Rudy Rucker's novel *Master of Space and Time*, the protagonists, Harry and Fletch, who have been using their time travel device to go back and forth among alternate worlds, suddenly enter a scene altogether outside the cosmos, or outside any possible world-lines. Harry explains: "We're in a three-dimensional cross section of infinite-dimensional superspace," in which the entire universe they initially inhabited is now represented quite straightforwardly as a small object, an "egg-shaped blob" of hazy white.[56] Within this superspace, Harry and Fletch are able to manipulate their own universe, as well as any other universe, essentially by retelling the stories in which they themselves are participating, standing outside them while explicitly occupying a concrete viewing-point, and then visually representing them as having such-and-such size, attributes, order, or duration. Fletch says:

> Looking more closely, I could see that our universe was really made up of a single tangled thread, a bright line that wove forward and back and in and out. It was like an endlessly knotted wire, a tangle of yarn, the Gordian knot. I looked at some of the other universes, knotty eggs all around us. We were really behind the scenes.[57]

This "infinite-dimensional superspace," or what mathematicians call a "Hilbert space," *is* a real extrapolation of mathematical conditions, and is in fact regularly invoked in the construction of physical models of the universe or of multiple universes. Theoretical physics has long dealt with hypotheses of physical scenarios that resemble Rucker's, Leiber's, or Williamson's; the current term for such a scenario is "metaverse," or sometimes "multiverse," an "ensemble of universes," or what Paul Davies calls a "vastly larger assemblage of 'universes,' or cosmic regions."[58]

The mathematical-physical prototype for such models is again Everett's Ph.D. thesis, "The Theory of the Universal Wave Function," which, as I mentioned in the Introduction, is far better known by physicists and *Star Trek* fans alike by its colloquial name, the "many-worlds interpretation" of quantum mechanics.[59] For present purposes, what is illuminating about Everett's model is that it engages in a materialist methodological rigor similar to Einstein's. Everett observes, as do all quantum theorists, that the necessary intervention of a physical measuring device, for instance a mechanical clock or a rigid rod, has the effect of rendering any given observation or measurement fundamentally indeterminate. We cannot predict a single value from the physical conditions of the system we try to measure, but only a range of possible values. The classic illustration of the problem Everett describes is Schrödinger's cat-in-the-box example, in which the enclosed cat is either poisoned or still alive, depending on whether a switch controlled by a quantum-scale event has been tripped or not. Prior to one's actually removing the cover and looking in the box, the cat's status is, as it were, 50 percent alive and 50 percent dead, but when one opens the box, one effectively "measures" its state to be one or the other; it cannot be both. The canonical explanation of why indeterminate values turn out to be determinate once they are measured—why there will always have been, after all, a definite value, a live or a dead cat in the box—is the "Copenhagen interpretation," proposed by Niels Bohr and Werner Heisenberg. The indeterminacy of the system "collapses" onto a single value the moment the act of measurement takes place. However, this explanation is widely thought to be not a physical but a metaphysical interpretation of the system and its measurement: in the mathematics itself, or in what physicists call the "formalism" describing the system, there is no immediate reason why the cat is not *both* alive and dead in the box, 50 percent each. All we have is a probability among an array of possibilities.

It would be feasible to say that the cat-in-the-box example appears, in its anecdotal, narrative form, to posit a viewpoint that its mathematical formalism strictly excludes. The Copenhagen interpretation shuts down that viewpoint only at the moment of measurement: the box is opened, and a metaphysical gesture now returns us, uncomfortably artificially, from a "later" moment in the story of the cat, in which a viewpoint is implied (I open the box), to an "earlier" one in which such a viewpoint was impossible but in retrospect already irrelevant (the cat is/was [already] alive/dead). Everett's model forestalls the uncomfortable *ex post facto* feel of that act of measurement, or of its "collapse," by sticking to the formalism alone, wherein no viewpoint outside the specific "universes" of the example—for instance, a viewpoint from which one could say either that the cat is definitely one or the other, *or* that the cat is both/neither—is ever implied.

Everett, then, in essence pursuing Einstein's empiricist rigor fully consistently, interprets his mathematics strictly literally, declining to append any nonphysical or metaphysical hypothesis to the story of the cat (or, more precisely, to the story of the probable states of the system): the cat is alive in one possible universe, dead in another, and each such universe must be considered real according to the underlying math. This goes for any indeterminate event, which is to say any quantum-scale interaction at all; here again is the physicist Bryce DeWitt, explaining the implications:

> This universe is constantly splitting into a stupendous number of branches, all resulting from the measurementlike interactions between its myriads of components. Moreover, every quantum transition taking place on every star, in every galaxy, in every remote corner of the universe is splitting our local world on earth into myriads of copies of itself.[60]

One might ask DeWitt, from what *viewpoint* is this brief passage about "myriads of [world] copies" narrated? I ask this question, not in order to impugn what is after all DeWitt's good-faith attempt to *narrate* mathematical and set-theoretical conclusions for a lay reader, but rather in order to help us see what the rigorous application of an antimetaphysical methodology like Everett's might offer us on the way toward theorizing a hyperspacetime viewpoint for *any* model of universes, including narrative ones. The seduction or compulsion toward a viewpoint is seemingly irresistible once one shifts from the formalism to its exposition, quite despite any caveat that no such viewpoint is physical. Where is DeWitt standing, so to speak, when he undertakes this vast metaversal reflexion about "stupendous number[s] of branches"? Is he in the same "Place" as Greta Forzane the prostitute or Asimov's kettle rider or the reader of *Madame Bovary*? Again, the time travel story, whatever else it may offer, is a superb means for depicting the position DeWitt adopts or assumes, as well as for working through the seemingly transcendental necessity of occupying such a position in order to narrate any relationships of factual and counterfactual possibility. For we are, with DeWitt, radically, as it were *infinitely*, "behind the scenes," as Rudy Rucker's character comments, viewing multiple or even all universes from superspace, even as we may strive to redress this epistemological hubris with an *ex post facto* declaration of its strict unphysicality. We are "on the Big Time"—and may now observe how utterly easy it is to be here, despite how ontologically outrageous our position would seem to be.

Moreover, it is not just easy, but metaphysically, psychologically, possibly sociologically compelling, even necessary, to occupy the impossible, unphysical space from which worldlines are reviewed and five-dimensional cross-spacetime stories are retold, once we leave behind the unnarratable register of mathematical formalism. The viewpoint of the time traveler, who comes conveniently before or after (or both) the specific narrative lines that compose a plot about any world or worlds, may be a transcendentally basic condition for speculating not only beyond, but *about*, the formalism of any real or ostensible physical world. I have inquired of DeWitt, above, where he presumes to locate the viewpoint from which he observes the "myriad" worlds entailed by Everett's interpretation of quantum formalism. Let me pose a similar query to Einstein, whose very careful proscriptions of nonphysical measurement ought to exclude the possibility of such purely "geometric" pseudospaces, along with any actions (measurements) undertaken within them. These would be the sorts of nonphysical actions and measurements performed by the entirely hypothetical speculator who schemes to peek inside Schrödinger's box in advance, so to speak, and view the cat before any physical measurement of its state is done. Such a speculator, whom it is certainly easy enough to imagine or characterize once the anecdote gets started, aspires to precisely the sort of unmeasured knowledge that the formalism of quantum theory directly prohibits, and achieves that knowledge through something like a time travel narrative, journeying into the physically nonexistent superspace of the unopened box. Like the uncannily metaphysical power of a movie camera—I will have more to say about such cinematographic viewing in the final chapter—this speculator is free to roam (at least in terms of the anecdote) where no strictly physical or mathematical agent ever could.

Einstein: Narratology of "Reality"

In a letter to Schrödinger in 1939, discussing this same cat-in-the-box ex-ample, Einstein writes, "I am as convinced as ever that the wave represen-tation of matter is an incomplete representation of the state of affairs, no matter how practically useful it has proved itself to be."[61] Einstein elabo-rates, now referring to two competing philosophical registers in current quantum theory upon which the ambiguity in the mathematics (Ψ-function) of the cat's state might be resolved:

> Both points of view [that the Ψ-function is either ontologically or only epistemologically ambiguous] are logically unobjectionable; but I cannot believe that either of these viewpoints will finally be established.
>
> There is also the mystic, who forbids, as being unscientific, an inquiry about something that exists independently of whether or not it is observed, i.e. the question as to whether or not the cat is alive at a particular instant before an observation is made (Bohr). Then both interpretations fuse into a gentle fog, in which I feel no better than I do in either of the previously mentioned interpretations, which do take a position with respect to the concept of reality.
>
> Of course I admit that such a complete description would not be observable in its entirety in the individual case, but from a rational point of view one also could not require this.[62]

Einstein here comes close to outright rejecting the very same "heuristic" positivism that grounded and animated his original discussions of relativity theory from 1905 through 1916. A conception of quantum events that are "not observable" and yet positable "from a rational point of view" seems perilously close to implying, for instance, the notion of simultaneity inde-pendent of any Galilean frame of reference, or a viewpoint from (unphysi-cal) superspace.

More than that, it appears that Einstein now accuses Niels Bohr of be-ing a "mystic" for holding something very much like Einstein's *own* earlier view, a view that excludes any inquiry into "something that exists indepen-dently of whether or not it is observed." Einstein thus not only adopts the position of a rationalist metaphysician that he seemed utterly to reject in his principal work on relativity and geometry, but, in what seems like a well-nigh dialectical reversal, accuses the ostensible positivist Bohr of "mystic[ism]" for "forbid[ding]" the theorization of physically unmeasurable phenomena, or for forbidding a viewpoint that could observe them, a proscription that is precisely the philosophical foundation of Einstein's original formulations of special relativity.

I feel far from qualified to speculate on Einstein's philosophical inconsistencies, but I do wish to note how seemingly *natural* it appears for the later Einstein, in correspondence with his fellow quantum skeptic Schrödinger, to adopt a "rational point of view"—a view that allows what appear to be strictly unphysical perspectives to enter into the theorization of physical events, and furthermore a view which suggests that events not "observable in their entirety" ought nevertheless to be permitted within a "complete description."[63] Here, Einstein does not carefully distinguish between the narrative superspace he occupies in order to speculate on the premeasurement condition of the cat in the box, and the physical theory one could develop strictly based on actual observations (measurements). In brief, I would identify this moment in Einstein's thinking as a fundamentally narratological one. He is now telling a time travel story: like Williamson, Leiber, Rucker, or even Flaubert, he narrates, from hyperspacetime, the superspace of possible narrative worlds, and only within that depicted scenography is he able to imagine that we might eventually learn how to go back into the box and find out whether Schrödinger's cat is alive or dead.

I wish to conclude this chapter by coming back to what is (for me anyway) the far more comfortable territory of a single time travel narrative, and by examining the complication it offers to this theorization of "the Big Time" of reading. Returning to Rudy Rucker's *Master of Space and Time*: after Harry and Fletch have been indulging for a while the quasi-theatrical fantasy of superspace, with its godlike directorial control over all possible universes and their sequences, Rucker suddenly interjects the thoroughly narratological idea that any such viewpoint, however seemingly ultimate and omniscient, must always require a *further* moment of narrative transcendence. In other words—and here I take Rucker to be extrapolating the implications of his physical description rigorously, like the earlier Einstein, or like Everett—there is no perspective fully outside of yet *another* potential narrativization, no perspective that only tells, and cannot be told and retold, or in short "re-flected," once we get started. Fletch suddenly says, "I kept having the feeling that we were being watched by some cool, detached intelligence just out of sight."[64] Rucker thus pursues a physical implication of the unphysicality of viewpoint that ends up looking uncannily Lacanian: if one is able to watch, then by necessity, even by definition, one can also *be* watched—if one reads, then one can *be* read and reread—and because this is the superspace of all possible universes, one *must* be. "Surely," Fletch says, "Harry and I were not the only beings to have entered Superspace."[65] That is quite right: so who else is there? The chapter title provides an answer: "Rudy Rucker is watching you."

Here, in a kind of culmination of viewpoint theory, the time travel story again literalizes a fundamental narratological condition, but now problematizes it to an extreme. First of all, Rucker's penultimate chapter demonstrates that there is always a transcendental position occupied in reading the possible world of a narrative, but that this position is never itself transcendent. Why? We have already observed that the transcendental space of reflexion, when depicted physically as a hyperspacetime or "Big Time" or superspace, ends up being structured just like the possible world of a narrative itself. They may be distinguished ontologically but they are not distinguished practically, if one can speak this way; each entails a structurally equivalent viewpoint. So if "the Big Time" behaves just like a regular narrative world, then it too entails or implies another space of reflexion. Or to put it a different way, as much as hyperspacetime *is* the space of reflexion, it is simultaneously also merely a narrative hypothesis to be further reflected; this further transcendence is required to tell or retell that narrative hypothesis, as with all narrative spaces. Therefore there will always be another "presence" watching, another viewpoint from another (hyper)spacetime. And in the end, we do not have a strict ontological basis for distinguishing the real world of reading, the narrative space in which we negotiate the relationship between possible worlds, from narratives themselves, the fictions we read or watch. Our world, too—the domain of reflexion, the putatively "real," "historical," "nonfictional" world—is also a *story*, one that must continually be retold, reassembled, *reflected*, in some ambiguously transcendental hyperspacetime, after the fact. It's turtles all the way down, or viewpoints all the way up.[66]

So, Rudy Rucker is watching you. But that is *in* the plot; in yet another superspace—and note how bizarre it sounds to say this, because superspace is the "infinite-dimensional" Hilbert space of all possibilities—in yet another superspace, Rudy Rucker is watching Rudy Rucker, ad infinitum. In rigorously physical-narratological terms, someone else is always watching, always reflecting, until, as a mere practical matter, we finally run out of narratives to tell or further reflexions to undertake.

Paradox and Paratext: Picturing Narrative Theory

Samuel R. Delany's 1966 novel *Empire Star* opens with this description of its protagonist, Comet Jo:

> He had:
>   a waist-length braid of blond hair;
>   a body that was brown and slim and looked like a cat's, they said, when he curled up, half-asleep, in the flicker of the Field Keeper's fire at New Cycle;
>   an ocarina;
>   a pair of black boots and a pair of black gloves with which he could climb walls and across ceilings; . . .
>   a propensity for wandering away from the Home Caves to look at the stars, which had gotten him in trouble at least four times in the past month, and in the past fourteen years had earned him the sobriquet, Comet Jo;
>   an uncle named Clemence whom he disliked.
>   Later, when he had lost all but, miraculously, the ocarina, he thought about these things and what they had meant to him, and how much they defined his youth, and how poorly they had prepared him for maturity.[1]

This catalog of traits and objects is fairly typical of popular science fiction of the period, offering a vocabulary dotted with exotic terms yet familiar in its overall generic pattern; it accords with what Delany himself, in a well-known essay, calls the "particular subjunctive level of SF."[2] More specifically characteristic of Delany, the passage begins to introduce hints of narrative complication—anticipated, in passing, by the ambiguity between the present "he had" and the future perfect "later when"—pushing what might otherwise have been a straightforward science fiction or fantasy tale toward something more like a nascent experiment in the shifting chronologies and chronometries of storytelling.[3]

A few pages into the novel, well before we are certain that this will be a time travel paradox story, Comet Jo encounters what appears to be a damaged avatar of his future self, come back to deliver a message:

> It staggered forward, smoking. It raised grey eyes. Long wheat-colored hair caught on a breeze and blew back from its shoulders, as, for a moment, it moved with a certain catlike grace. Then it fell forward.
>
> Something under fear made Comet Jo reach out and catch its extended arms. Hand caught claw. Claw caught hand. It was only when Comet Jo was kneeling and the figure was panting in his arms that he realized it was his double. . . .
>
> . . . The accent was the clean, precise tone of off-worlders' Interling.
> "You have to take a message to Empire Star!"[4]

As Delany's novel unfolds, the reader attempts to track the logic of such doublings and reversals, finding them recurring increasingly and uncannily as the story approaches its end. By that point, the plot has begun directly to repeat and recontextualize some of its own earlier events, creating extreme versions of what Genette calls "anachronies," temporal disjunctions between what is happening in the plot and what is happening in the underlying events it represents.[5] For instance, earlier in the novel Comet Jo is mentored by an older and more experienced space traveler, San Severina, who gives him a red comb. Much later, the adult Comet Jo again encounters San Severina, who is "now" younger than before, indeed a child. Jo becomes *her* mentor, alerting her that she will eventually encounter him in an "earlier" version:

> "[A] long time from now, while what I am telling you about now is happening, you will run into a boy." Jo glanced at his reflection on the glass. "I was going

to say that he looks like me. But he doesn't, not that much. . . . He'll be a lot browner than I am because he's spent more time outside than I have recently. His speech will be almost unintelligible. Though his hair will be about the color of mine, it will be much longer and a mess—[.]" Suddenly Jo reached for his pouch and dug inside. "Here. Keep this, until you meet him. Then give it to him." He handed her the red comb.[6]

The comb that Jo gives San Severina is the same one that she had earlier given him; each character is both mentor and protégé, in a chronologically and causally irreconcilable juxtaposition of sequences that nonetheless forms a single coherent plot. In other words, we now have a typical time travel paradox, a set of events or lines that will not resolve themselves into a single chronological or causal sequence. Shortly I will examine what might be considerably less typical about Delany's approach to such paradoxes, and what therefore makes his novel especially useful for sorting out epistemological issues latent in the construction of narrative viewpoint, issues that paradox stories thrust into a crisis. For now, however, we are playfully running the kind of relatively mundane narratological obstacle course that Stanislaw Lem calls "a perfectly ordinary time loop."[7]

In my Introduction, analyzing a scene from Silverberg's *Up the Line*, I suggested that we might interrogate the epistemology of time travel fiction by posing straightforward questions about chronology, and then observing whether straightforward answers to such questions, of the sort we could expect with respect to "classical" narratives,[8] are forthcoming. *Empire Star* invites the same sort of disingenuously straightforward chronometric inquiry. One might ask, simply, who possesses the comb *first*, Comet Jo or San Severina? A preliminary answer to this question would be that in the book's early pages the comb is in the possession of San Severina; later on she hands it over to Comet Jo, and still later he returns it to her. Such an answer refers us only to the order of comb exchanges as the author contrives to arrange it in the novel's distinct sequence—this is the order of the "plot" or "discourse," which for the most part we can track conveniently in the physical text: chapter 4 or 11, page 28 or 85, and so on. Of course, such an answer is unlikely to be satisfying if we are inquiring about what is earlier or later in the actual story as the characters experience it within their own world. The ordering of pages or chapters in the text is after all merely a convenient collation for the reader, not for them—but really also not for us, to the degree that we "enter" or "travel to" the story world. So, for instance, if someone inquiring about the story of Charles Foster Kane were to ask, when does Kane utter the word "rosebud," one could answer either: at the beginning of the movie *Citizen Kane*, or, at the end of Kane's life. Only the latter answer contains the crucial biographical information that "rosebud" is Kane's last word regardless of its appearance in the film's opening scene. Without this basic grasp of the referential lastness of the word "rosebud," the story loses its principal pathos, both formal and psychological. Similarly, in *Empire Star*, we require an answer that tells us who had the comb first in the underlying story. We are asking which character "really" had the comb first, or "in fact" had it, in a life, not in a chapter, entirely in spite of the discursive order in which Delany elects to place these events in the published book. As I will show, a paradox story offers a fruitfully provocative non-answer to such a question, opening up a certain theoretical crisis at the points where *fabula* and *sjuzhet*—the discourse and the life story of the narrative—strive for reconciliation, and even at the points where the book itself appears to re-present such a reconciliation.

The Postulate of Fabular Apriority

Virtually all narratological theories identify multiple levels or "layers" within narratives,[9] usually predicated upon a conventional binary, the consensus surrounding which Jonathan Culler describes as follows:

> [I]f these theorists agree on anything it is this: that the theory of narrative requires a distinction between what I shall call "story"—a sequence of actions or events, conceived as independent of their manifestation in discourse—and what I shall call "discourse," the discursive presentation or narration of events.[10]

Culler's particular terminology is lifted from Seymour Chatman, who pithily glosses the pairing Culler identifies: "I posit a *what* and a *way*. The what of narrative I call its 'story'; the way I call its 'discourse.'"[11] H. Porter Abbott, like Culler, calls this "analytically powerful distinction" the "founding insight of the field of narratology."[12] Indeed, for narrative theorists following formalist or structuralist legacies, some version of the binary is all but axiomatic.

But, as Abbott almost immediately observes, "nothing is tidy in the theory of narrative," and theorists are rarely as content as Chatman either with the specific terms or with a finite number of them; complications and multiplications abound. To give just one influential example, Mieke Bal uses "story" for Chatman's "discourse" and "*fabula*" for his "story," while adding the additional narrative layer of "text." Most theorists, in fact, add one or more layers beyond the pairing of story and discourse. A certain triple schema achieves a somewhat less easy consensus than the simple story/discourse binary—or at least an agreement upon where to disagree—around Gérard Genette's tripartite usage in his canonical *Narrative Discourse*: *histoire* (story), *récit* (presentation of events in discourse), and *narration* (narrating or enunciation).[13] The supplementary terms or layers are motivated in part by the difficulty of ascertaining how "discourse" could remain a coherent figuration for representing "story" without reference to a material or quasi-material substratum in an actually produced text. Yet even this tripartite or multiple usage tends to devolve back into the more basic or convenient binarism that Culler indicates, and that was first identified by the Russian formalists as *fabula* (story material) and *sjuzhet* (narrative re-presentation; plot). Genette himself, as Bal observes, "in the end . . . distinguishes only two levels, those of Russian Formalism."[14] These two levels, then—largely for the practical purpose of reconnecting the terminology to its conventional roots and providing its most primal version—are the ones I will provisionally retain for my discussion, adopting their names directly from the foundational terminology of the Russian formalist critic Viktor Shklovsky: *fabula* and *sjuzhet*.[15] Ultimately, I too will propose an additional necessary level—"paratext," as it turns out, rather than "text," for reasons I will discuss shortly.

I have already noted, for example with respect to *Empire Star*, that authors and readers do not treat the *fabula/sjuzhet* binary as symmetrical. The most salient feature of Shklovsky's originary pairing, for the purposes of understanding ordering in narratives, is that it is a hierarchical arrangement and stridently assumes the temporal and causal priority of one of its poles: *fabula* appears to come *first*. This is in one sense surprising vis-à-vis fictions, because clearly both *fabula* and *sjuzhet* are equally constructed or posterior in the writing and reading of, say, a novel; strictly speaking, all levels of narrative literature are concurrently mediated. As Monika Fludernik notes, "The author of a novel or a film script develops a fictional world and produces both the story and the narrative discourse that goes with its product, the narrative text."[16] The self-evidently simultaneous double production of *sjuzhets* and *fabulas* leads Fludernik to assert that "fictional narrative, whether in fairy tale, novel, or television film, differs radically from historical writing"; in essence, "fictional narratives create fictional worlds, whereas historians collectively seek to represent one and the same real world."[17] If *fabula* still appears to come first in a novel or film, this appearance results only from the general constellation of psychological and performative acts by which we "suspend disbelief" in our encounter with fictions, merely pretending or agreeing to behave as if texts are referring to prior events or facts.

Yet, despite Fludernik's perhaps sanguine claim for this "radical" difference between fiction and history, most of twentieth-century narrative theory, if not exactly conflating the two categories, has largely behaved as if the correspondence between fiction and history were troublingly implacable, something akin to an ideological belief or a phenomenological apperception on the part of readers or writers, and therefore a continuous occasion for critical qualification or refutation. At least as far back as Henry James, theorists have observed that "story"—which is to say, the register or layer of narrative that appears to occupy the role of the referent, the ostensible "material" being reconstructed in a new present[18]—operates in many ways just like history, or occupies the structural position of a factual record. During an extended critique of Anthony Trollope, James writes that the novelist's vocation or, more precisely, the novelist's posture or pretense, is therefore exactly that of a historian:

> It is impossible to imagine what a novelist takes himself to be unless he regards himself as an historian and his narrative as a history. It is only as an historian that he has the smallest *locus standi*. As a narrator of fictitious events he is nowhere; to insert into his attempt a back-bone of logic, he must relate events that are assumed to be real. This assumption permeates, animates all the work of the most solid storytellers.[19]

The technical or formal distinction between historical events "assumed to be real" and their (subsequent) narration in a plot, a distinction to which the broader generic differences between fiction and nonfiction is sacrificed, clearly entails a differentiated hierarchy like that devised by the Russian formalists, the *fabula/sjuzhet* pair. Moreover, James's description captures a sense of the more ancient notion of "suspension of disbelief" that still confronts or confounds narrative theory, the notion that for identification or imagination to occur in a narrative—for the story to *compel*—readers must take a fiction to be true, must enter the world of story or, on the flip side, must temporarily or hypothetically exit the world of the actual produced text. Following simply on the fact that James is able to hypothesize such an equivalence between fiction and nonfiction at all, I would go as far as to assert, now directly against the spirit of Fludernik's "radical difference," that the ostensible historicity of even a fictional *fabula* might be said to constitute a cherished or fetishized axiom for narrative theory, possibly for the very act of reading itself.[20] This is the case even though certainly there has never been a sane narrative theorist (I presume not even James, when it comes down to it) who would share such an absolute insistence that the novelist act like a historian, or that "as a narrator of fictitious events he is nowhere," a stance that J. Hillis Miller calls James's "hyperbolic" defense of the author's pretense. It might then be plausible to characterize the entirety of twentieth- and twenty-first-century narrative theory as the continuous history of reactions *against* its own axiomatic ground, most directly articulated by James. Thus Tzvetan Todorov responds by calling this axiom the "representative illusion"; Genette, the "referential illusion"; and Peter Brooks, the "mimetic illusion."[21] In each case, what is demystified or unreified by the critic is a bluntly incorrect but nonetheless persistent presumption that *fabula* is prior, that it contains "'what really happened'";[22] the plot merely retells it.

The power of such an "illusion" may be gauged by the lengths and depths to which theorists go in order to repudiate it. When Todorov insists that in a piece of literature "there is not, *first of all*, a certain reality, *and afterward*, its representation by the text," or when Wolfgang Iser likewise insists that "the text is in no way confined to being the representation of something given,"[23] clearly an elemental premise is that on some precritical register of reading the *fabula is* "assume[d] to be real" (James), and therefore necessarily a target for critical *intervention*, almost in the pop-psychological sense of the term. It therefore must also be the case that readers set about to "naturalize" a narrative as they receive it, as Chatman argues following Culler, "'forget[ting]' its conventional character" and adjusting it to "facts and probabilities in the real world"—agreeing, somehow, to act as though the *fabula* is prior fact or true history.[24]

In the understanding of fictionmaking implied by these critics, fictions perpetrate a peculiar ruse upon the ostensibly real or historical world, a kind of highly conventionalized lying. Lubomír Doležel, in his detailed advocation of a "possible-worlds" model of fictionality, outlines a common theory of what he calls the "pseudomimesis" of fiction, which "presuppose[s] that fictional particulars somehow preexist the act of representation" and that authors then "report," "describe," or "share . . . knowledge" about such particulars.[25] For Doležel, the roots of a theory of pseudomimesis are traceable not to James but rather to Bertrand Russell, whose position Doležel describes as the "extremity of the one-world semantics of fictionality."[26] Russell's claim is that there is only one real world, and that therefore a fiction writer can at best pretend to refer to it, very much in the same way that James's novelist "takes himself to be" a historian; Russell goes as far as to describe fictional sentences merely as false or, more precisely, as "not denot[ing] anything."[27] Doležel proceeds to cite John Searle's entirely Russellian claim, which again echoes James closely: "'The author of a work of fiction pretends to perform a series of illocutionary acts,'" playing a "game" in which a persuasive reference to something putatively actual is the game's objective.[28] Thus Doležel identifies within this game an "ontological commitment": "an especially enthusiastic player might get so caught up in the linguistic game that it turns into an ontological one."[29]

It is just such a commitment to ruse, and the concession to its power in generic narrative fiction, that I will analyze in the time travel paradox story. I am less concerned with the ultimate accuracy of the term "ruse" (or "pseudomimesis") as a description of fictional construction than I am more broadly with the pull that a "one-world" historicism exerts upon readers, writers, and critics of fictions. For the moment, then, I wish to supply a term for the ruse that combines the hierarchical sense of the *fabula/sjuzhet* binary with the apparent phenomenological compulsion, in the reading act, toward a "one-world semantics" of fictionality. To some extent following James, I will suggest that fictions *postulate* their *fabulas* as "factual" and "prior" to their *sjuzhets*—upon this postulation rests any suspension of too-obvious fictionality required to keep narratives moving and developing. I will call the ruse of fictional historicality the "postulate of *fabular* apriority."

The Referential Pathos

A postulate of *fabular* apriority presumably underlies realism of all sorts, a mode of storytelling that straightforwardly maintains the "mimetic illusion" that the underlying *fabula* is self-consistent, potentially reconstructable, and prior to its ostensible retelling.[30] All shifts in narrative viewpoint, reversals of order, hiatuses, repetitions, and so on, therefore occur only on the level of *sjuzhet*, the layer that comes "after" *fabula*. Earlier I described this scenario, following Genette and Chatman, as "classical," a better term than "realistic" because it encompasses the wider range of stories and story types that postulate *fabular* apriority. In a fantastical story, for instance, *fabula* may diverge from real life in distinct nonrealistic ways—vampires exist, or witches cast spells, or there is a time and place called Middle Earth (or Discworld or New Crobuzon)—but the underlying story nonetheless remains coherent and regular, especially in terms of its chronological order; indeed, something akin to Kantian transcendental spacetime conditions remains solidly in place. The One Ring is first forged, then discovered by Gollum, then by Bilbo, then by Frodo, then ultimately disposed of, regardless of the flashbacks, cross-cutting of scenes, or delimitations of perspective employed by the narrator (or, alternatively, by Tolkien) to convey these sequences of events. Indeed, among other things, it is the coherence of temporal order in fantastic *fabulas* that yields the resemblance of romances of all sorts to "myths, folktales, fairytales—the prototypes of all narrative, [and] the ancestors and the models of later fictional developments."[31]

In other types of nonrealistic fiction, such as stories of the uncanny or the absurd, problems or ambiguities in a *sjuzhet* might hint at ambiguities of the *fabula*, for instance in the form of psychological or sociological vicissitudes. The pathos of a ghost story can rest on the question of whether uncanny events in the plot are to be resolved either via psyche (in terms of a skewed narrative perspective such as hallucination or dream) or via the supernatural (in terms of actually occurring mystical events). The theorist's prototype for such ambiguity is often James's *The Turn of the Screw*, in which the first-person narration of the governess, a dubious witness at best, famously leaves uncertain the authenticity of ghostly apparitions in the plot. In brief, their super-nature is indeterminable in the *sjuzhet*—I surmise this indeterminability not strictly from the story itself but also from its long afterlife as a source of useful trouble for James's critics—since the reader, who is consistently "focalized"[32] through the (possibly) unhinged governess, has no independent access to the *fabula* sufficient to fix an empirical or causal interpretation. Nonetheless, the story proceeds with, even relies upon, the assumption that such facts *would* be discoverable by an independent observer, were one only available, and that the story's world is therefore complete despite the lacunae in its *sjuzhet*.[33] Indeed, the effectiveness of any unreliable narrator is grounded in precisely the same historicist postulation of *fabular* apriority that James insists upon for strictly realist fiction. The power or interest of such a narrator is in his or her specific divergence from *fabula*, and therefore the classical coherence of *fabula* is still presumed in principle, even if inaccessible in practice.

Any brief survey of unreliable narrators furnishes a variety of ways in which *fabula* might be fruitfully obscured by *sjuzhet* yet still postulated as prior: psychologically troubled narrators (Kingsley Amis's *The Green Man*, Bret Easton Ellis's *American Psycho*), narrators inhibited by bias or intense self-absorption (J. M. Coetzee's *Waiting for the Barbarians*, Chang-Rae Lee's *A Gesture Life*), narrators focalized by deep-rooted social habitus or trauma (Kazuo Ishiguro's *When We Were Orphans*, Toni Morrison's *Beloved*, Nicole Krauss's *The History of Love*), narrators whose naïveté or social incapacity inhibits insight (Emily Brontë's *Wuthering Heights*, Ford Madox Ford's *The Good Soldier*), or narrators who, overpowered by events or self-interest, construct multiple or incompatible *sjuzhets* (Wilkie Collins's *Moonstone* and *The Woman in White*, Akira Kurosawa's *Rashomon*, William Faulkner's *As I Lay Dying*). In all these cases, despite even radical revisionism or experimentation on the level of *sjuzhet*, the *fabula* remains classical, which is to say, "historically" linear and causal, and potentially reconstitutable as a past.

In self-conscious or metanarrative fictions, for instance in the blatant fictionalizing about which James complains in his reading of Trollope, the "mimetic illusion" might be explicitly problematized or even shattered. Metanarratives, still radical perhaps even through a good part of twentieth-century modernism, have by now become sufficiently commonplace that we may actually be in the habit of assuming that the classical form of the *sjuzhet/fabula* relation, with its maintenance of the postulate of *fabular* apriority, has been permanently subverted and is finally obsolete. My five-year-old daughter watches a television show called *Word Girl*, a Trollopian metafiction in which the voiceover narrator converses directly with both the characters and the viewer about the progress of the episode, its relation to other episodes, the length of the show itself, the repetition of conventional motifs, and the implausibility or predictability of plot coincidences. Yet even here, what lends the narrative its pathos—even as an experiment in recasting or exposing the five-year-old viewer's own classical cathexis upon the plot, his or her suspension of disbelief in its fictionality—is precisely that the *fabula* retains its classical *function* as referent, while the *sjuzhet* foregrounds its own constructive role as a deliberate project of signifying and resignifying. My daughter continues to understand that all variations in storytelling are *sjuzhet*-bound, that (to the degree a children's cartoon can be said to assume this) the lives of the characters and their relationships through time proceed apace in the background, postulated as a history to which the narration continually refers, even as it may also reference itself. Indeed, the pathos of any self-conscious narrative or metanarrative is precisely still the originally unselfconscious pastness or truth of *fabula*, its primal historicity, now exposed for scrutiny or parody and possibly destroyed, as James feared in Trollope's writing or as we have come almost to presume in postmodern fiction, film, and television. We might go as far as to state that even instances in which *all* readerly efforts to reconstitute "historical" coherency or apriority are thwarted or fully ambiguated—Julio Cortazar's *Hopsotch*, Tom Tykwer's film *Run Lola Run*, hyperfictions such as Shelley Jackson's *Patchwork Girl*, or constrained formal exercises such as Ernest Vincent Wright's *Gadsby* or Walter Abish's *Alphabetical Africa*—retain the notion of classical or historical *fabula* as a realm superseded, and may therefore be considered dialectical revaluations of other versions of realism and antirealism, continuous, in an Aeurbachian sense, with the prior history of narrative fiction and its postulates. Even with these last examples, although they are definitively "meta-fictions," with a style of reference primarily or only to *sjuzhet*—Linda Hutcheon has called such writing a "mimesis of process"—the underlying classical *fabula* is still retained as a lost or outmoded pattern.[34] *Fabular* apriority may be conspicuously dissimulated or destroyed, yet it continues to lie at the now quasi-mythic origin of revisionist technique through the persistent structural or even ideological "nature" of its precedence.

Finally, then, I feel warranted in speaking of "referential pathos"[35] in narratives, analogous to the pathos of "return" that drives the psychoanalytic symptom. Narrators, even highly self-conscious or subversive ones, return to the *fabula* as though to the founding scenography of a retelling, sometimes explicitly, as in realist fictions, sometimes symptomatically, via multilayered renarrations of (the pathos of) a postulated past. Stories not only illustrate but also *compel* the postulation of *fabular* apriority; *fabula* is not merely prior, but *primal* for any given narrative, a founding fantasy. In other words, *fabular* apriority is a matter not so much of form as of desire, perhaps related to what Alain Badiou famously calls "the passion for the real".[36] something in the reader *wants* the narrative to be other than what it is (a fiction)—wants it to be actual, to *refer*. I might assert—tentatively, since this touches on topics too large for the present discussion—that only such a pathos, the psychologically primal facticity and priority of the *fabula*, could make plausible James's otherwise extreme or even fantasmatic claim that the novelist is a historian.

Indeed, a whole body of cognitive research sets out to demonstrate that the "suspension" that permits narrative involvement or "transport" may entail that we are inherently disposed to treat stories in the way that James wants us to, as historical pretense.[37] At the polar opposite of Fludernik's claim for a "radical difference," Richard Gerrig concludes that "there is no psychologically privileged category 'fiction.'"[38] And this is quite alongside the psychoanalytic models that critics such as Brooks or Shoshona Felman adopt, in which narratives are in part animated by the "need to restore an earlier state of things."[39] It is not by coincidence that the ambiguous relationship of *fabula* to *sjuzhet* closely resembles the ambiguous temporality of trauma and symptom noted by Freud in several case studies but elaborated most fully in "The Wolf Man."[40] In a famous moment of methodological candor,[41] Freud observes that the "primal scene" of parental coitus that he and the Wolf Man together discover at the root of the latter's obsessional symptoms—"the origin of all origins, the bedrock of the Wolf Man's case," or what Brooks calls "the 'ultimate' *fabula*"—might never have taken place, and might therefore exist only as a reconstructed fantasy projected backward in time.[42] What is primal about the primal scene is therefore only its *cathexis* as temporal. Even if it is empirically factual, or even if we could ever confirm this, nonetheless it is still a fantasy constructed from fragments of other, later observations, projected into the position of originary *fabula* for the etiological plot that the Wolf Man is in the process of constructing and recathecting. Thus Freud observes that in the end it does not matter whether the primal scene is fact or fiction, a historical event or a fantasmatic reconstruction: the patient's subsequent narrative is the same in either case. What remains primal in the symptom is only the ostensible historical or real character of its originary referent, as it were the feel or tone of historical reality: in short, its referential pathos, or the emotional attachment to its postulated apriority. Finally, the whole hoary question of "suspension of disbelief," as well as the question of "identification" and the even more ancient problem of "imagination" in reading—none of which are by any means superseded by contemporary narratological technics, however sophisticated—might be recast as a symptomatology of the fantasy of the primal *fabula*, the cathected postulate that *fabula* precedes *sjuzhet*, that plots are (re)told, but that their stories *are*, or *were*.

Multiplex Narration in Empire Star

From a theoretical perspective, time travel narratives have at least the following distinct advantage over classical narratives, even ones of an extreme metanarrative or antinarrative form: time travel depicts realistically an explicit crisis of *fabular* apriority and of its motivating drive, the referential pathos. Like a psychoanalytic case study, yet more straightforwardly or in more aesthetically efficient form, time travel paradox stories lay out the conditions and limitations of our "naturalized" postulation of *fabula* through the retrospective construction of a *sjuzhet*. And finally, time travel stories refer us to the place where the negotiation of *sjuzhet* and *fabula* begins, namely to the text—or rather, as I shall argue, to the paratext. But I return to Samuel Delany's *Empire Star* in order to scrutinize the construction of these conditions and limitations *in situ*.

We previously observed that an answer to our straightforward question about *Empire Star*—who first had the comb, Comet Jo or San Severina—

cannot be given in terms of "classical" reference to the *fabula*. Such incapacity is not merely a matter of specific knowledge—say, a lack of access to
information or an inadequate point of view—but rather is true in principle.
There is no single answer for this question in the story's ostensible "reality,"
or in the sequence of "facts" that are related or retold on the level of *sjuzhet*;
each character had the comb before the other; indeed, each had it before it
had ever been given. The time travel paradox, in this case the pair of mutually incompatible comb exchanges, at once demands and refuses reconstitution as "historical" sequence, while nonetheless resolving itself into a single
tale. Under what conditions, then, can it accomplish such a resolution?

Having established the terms of a paradox, which is to say an irrecoverable indeterminacy of *fabula*, Delany proceeds to introduce into the narration successive complications that help render the indeterminacy a workable
component of the plot. The very first such complication is a minor one: the
phrase "they said" in the third line of the passage quoted in my chapter's
opening, which constrains the narrative voice, or "focalizes" it, to use the
narratological term, suggesting a more restricted viewpoint than any purely
objective one—perhaps the viewpoint of a person privy to local gossip or
lore. Note that in the continuation of the passage, the phrase "until the
end," like "they said," connotes this new focus of the narrative's voice: "Before he began to lose, however, he gained: two things, which, along with
the ocarina, he kept until the end. One was a devil kitten named Di'k. The
other was me. I'm Jewel."[43] For Delany, such shifts from an objective to a
more focused or local narrative perspective are more than mere convention,
as the explicit disclosing of the narrator, Jewel, soon lets us know. Jewel
represents not just the selection of a first-person speaker from among the
possible speakers within this plot. Rather, Jewel is a theoretical agent, a
type of narrative viewpoint, explicitly described as belonging to the scene
of writing:

> I have a multiplex consciousness, which means I see things from different
> points of view. It's a function of the overtone series in the harmonic pattern of
> my internal structuring. So I'll tell a good deal of the story from the point of
> view called, in literary circles, the omniscient observer.[44]

Jewel now becomes the speaking position against which all other viewpoints in the book are measured and even critiqued. Specifically, Jewel's

"multiplex" consciousness is contrasted with a "simplex" consciousness that merely reacts to stimuli and information, and a "complex" consciousness that has the capacity to see multiple viewpoints but not yet to synthesize them into unified patterns. This tripartite schema becomes most theoretically fruitful toward the end of the novel, when Comet Jo himself has evolved from simplex to complex to multiplex, and Jewel is freed to indulge the fuller range of its interest in the ontological status of the narration, as one presumably does "in literary circles." I will look more closely at Jewel's newly freed speculation in a moment.

So far I am still describing classical fictionmaking, merely shifted from a more ostensibly naturalistic to a more self-conscious or metanarrative mode. Returning to the paradox of the comb exchanges, and to what I identified as the inexorably nonclassical incompatibility of *fabula* and *sjuzhet*, we can begin to observe the specific difference the presence of Jewel makes to the narrative and to the two scenes of comb giving. This can be done, appropriately enough for a time travel story, in terms of dimensions. In the "earlier" scene, the one in Delany's chapter 4, in which Comet Jo is given the comb by the older San Severina, a simplex consciousness, in the person of Jo himself, observes merely the single train of events along the line of *sjuzhet*. For Jo, the events unfold thermodynamically, so to speak: he goes about his business on a continuum from past to present, receiving the comb at a specific point, then possessing it through ensuing future points. Up to this stage, the chronology of *sjuzhet* strictly corresponds to that of *fabula*, and Jo lives his own narrative in a linear real time, with nothing to interfere with a "historical" reconstitution of the events transpiring.

A reader, of course, is never quite so blindered, and occupies something approaching a complex or bidimensional viewpoint, already observing that the simplex line of events proceeding from past to future will be inadequate to explain not only the perplexing comb exchanges and their significance but indeed the rest of Jo's life. Signs of the need for such a complex perspective, and ultimately a multiplex one, are offered by Jewel, who, as we gradually learn, has the capacity to range freely over timelines or worldlines, as well as in multiple directions, and accordingly manipulates both ellipsis and paralipsis in the unfolding *sjuzhet*.[45] In the following passage, Jewel does so in the characteristically literal mode in which time travel narratives permit their narrators to indulge, and which still takes advantage of, or jokes about, the referential pathos felt by the reader in reconstructing *fabular* sequence:

> I left the incident out because I thought it was distracting and assumed it was perfectly deducible from Jo's question what had happened, sure that the multiplex reader would supply it for himself. I have done this several times throughout the story.[46]

Jewel's position is somewhat like that of the *reader* of a tale like Heinlein's "By His Bootstraps" or Williamson's *The Legion of Time.* In order to reconstitute the plot properly into its single *fabular* line, to the degree that this can be done, such an agent must now acquire a narratological perspective capable of reconciling even potentially paradoxical changes in the lives of characters, in order eventually to render them equivalent, to permit the characters properly to "survive" the transformations they undergo in the *sjuzhet.* The reader must learn, in a sense, to trump all simplex alterations with a multiplex viewpoint that is able simultaneously to distinguish and to equate multiple lines within a narrative multiverse. Jo himself is beginning to acquire such a multiplex viewpoint as the story grows more dense:

> "I see. I see you and me and Di'k and Jewel, only all at the same time. And I see the military ship waiting for me, and even Prince Nactor. But the ship is a hundred and seventy miles away, and Di'k is behind me, and you're all around me, and Jewel's inside me, and I'm . . . not me anymore."[47]

Here, the perspective of a character within the plot itself evolves from its simplex immanence in the mere course of events into something approximating the multiversal perspective of the narrator or reader, a position not localizable within anything like a single subject—hence Jo's declaration "I'm not me anymore." In fact, Jo is becoming something less like a character and more like a proxy for the narrator, Jewel, therefore able to review critically his own prior postulations of *fabular* apriority, even to analyze his own referential pathos. Delany eventually concludes by telling us that the very old Jo becomes the figure "Norn," who, in the opening pages, accompanied by Jewel and taking the form of Jo's damaged double, had first given Jo the message for Empire Star. The story thus repeats itself in several circuits, as it were, and the reader is explicitly tasked with adopting a multiplex position, a "viewpoint-over-histories," sufficiently nuanced to sort out these orbits and vectors and to locate the limits of their coherence.[48] Elsewhere, Jewel states: "The multiplex reader has by now discovered that the story is much longer than she thinks, cyclic and self-illuminating. I must leave out a great deal; only order your perceptions multiplexually, and you will not miss the lacunae."[49]

This last quotation commences the final chapter of the novel, in which Jewel partially ceases to narrate and addresses the reader directly in that second-person mode Garrett Stewart calls "conscription."[50] The plot overall is then reviewed as a kind of visual tableau, sometimes figured by Jewel as a "mosaic" composed of "tiles," at other moments literally as a book with pages, to be re-sorted with varying degrees of skill by its potential readers:

> *No end at all!* I hear from one complex voice.
> Unfair. Look at the second page. There I told you that there was an end and that D'ik, myself, and the ocarina were with him till then.
> A tile for the mosaic?
> Here's a piece. The end came sometime after San Severina (after many trips through the gap). . . . [51]

It now becomes very clear that Jewel's *outré* epistemological achievement, the multiplex assimilation of mutually incompatible *fabular* lines, is a duplication of the reader's own repertoire of perspectives, as well as of the metalinear position into which he or she is finally driven by the referential pathos. The time travel paradox story itself becomes, almost fully literally, an illustration of the practical limits of "normal" storytelling, or an illustration of what is required to remake the abnormal into the normal, to re-postulate *fabular* apriority in the face of its paradoxical impossibility. Jewel's earlier equation between the abstruse construction of its own viewpoint, "the overtone series in the harmonic pattern of my internal structuring," and the conventional "point of view known, in literary circles, as the omniscient observer," becomes fully tangible as the reader's own synthetic viewpoint-over-histories. The paradoxical *fabula* has compelled the reader's multiplexity, not necessarily as some highly sophisticated experiment in hyper- or metafictional reading, but rather as a basic technical means of reconstituting *fabula* at all.

In keeping with the tongue-in-cheek rendering of the reader's proxy viewpoint as a quasi-science-fictional technology (an "overtone series," etc.),

Delany peppers the novel with explicit metatextual references, recalling to the reader's mind the *sjuzhet* that is unfolding or proliferating, both in its multiplexity and in its teleological momentum toward full comprehensibility. "[T]hink how tiring your clumsy speech will be to your readers," San Severina says to the still-simplex Jo; "If you don't improve your diction, you will lose your entire audience before page forty."[52] Elsewhere the cyborg character Lump (an acronym for "linguistic ubiquitous multiplex") teases Jo using the names "Oscar," "Alfred," and "Bosie," a literary-historical reference perhaps unlikely to be immediately recognizable to the novel's science fiction readership: Oscar Wilde's love affair with Lord Alfred "Bosie" Douglas. "You didn't get it?" Lump asks Jo; "It was a literary allusion. I make them all the time."[53] Curiously, such high-minded literariness becomes one way for Delany to draw the reader's attention back to the political status of the novel, never fully elaborated but always present at the fringes of its narrative recomplications.[54] The novel's multiplanetary society relies upon a race of slaves, the Lll, who "built over half the Empire" and who now, through powerful extra-sensory emanations, imbue everyone who comes near them with "an overpowering sadness."[55] "No man can be free until they are free," San Severina says, echoing a line commonly attributed to Martin Luther King (and sometimes to Gandhi) but more directly present in the cultural milieu of 1965 via John F. Kennedy's famous Berlin speech of two years earlier: "[W]hen one man is enslaved, all are not free."[56] With racism, slavery, and imperialistic expansion weaving their way through the novel's plot, Lump and Jo exchange light jokes about the generic narrative features that gradually expose these weighty topics to the reader's attention:

> "You're a Lll?" Jo asked again, incredulously. "It never occurred to me. Now that you've told me, do you think it will make any difference?"
>
> "I doubt it," the Lump said. "But if you say anything about some of your best friends, I will lose a great deal of respect for you."
>
> "What about my best friends?" Jo asked.
>
> "Another allusion. It's all just as well you didn't get it."[57]

A short while later, Lump asserts that he refuses to "pass" as a noncomputer, again just flashing hints of a racial motif.[58]

We see, then, that not only temporal manipulations, but also the political and ethical subplots that animate and nuance Comet Jo's journey of self-discovery, are made available to the reader as potential threads of a larger multiplex fabric. But they are made available in the mode of metatextual gibes or jokes, reminders of the *manner* in which the *sjuzhet* is in the process of arranging and rearranging the various strands of *fabula*. Such allusions—both intra- and intertextual—have the cumulative effect of concretizing the reader's position, rendering it as a plotlike site of direct address, indeed something like a character who must periodically relearn his or her multiplex role in collusion with the increasingly self-assertive narrator: "Remember me? I'm Jewel"; "I hope you haven't forgotten me, because the rest of the story is going to be incomprehensible if you have"; "I'm Jewel."[59]

Toward the end, Delany finally connects the novel's political stakes directly to its paradoxical structuring. Lump, accompanied by the now young San Severina, explains to Jo (again with an intertextual allusion) that Empire Star is a "still point in the turning world":[60]

> "The fibers of reality are parted there. The temporal present joins the spatial past there with the possible future, and they get totally mixed up. Only the most multiplex of minds can go there and find their way out again the same way they went in." . . .
>
> "It's a temporal and spatial gap," the girl explained. "The council controls it, and that's how it keeps its power. I mean, if you can go into the future to see what's going to happen, then go into the past to make sure it happens like you want it, then you've just about got the universe in your pocket, more or less."[61]

Thus, amid this mélange of political allusions, literary self-references, and viewpoint repositionings, Delany's narrative eventually reassembles a *fabula* that, although ultimately still ambiguous in terms of temporal order, possesses sufficient dialectical elasticity to permit the adjudication of paradoxical lines in the form of a juxtaposition of mere incompatible alternatives, for instance in the form of a virtual map or diagram—in short, a time travel paradox tale. And having acquired such a virtual diagram, we also arrive at the second comb exchange, as Jo realizes who he is (or was) and who San Severina is (or was) and achieves what the reader too has been striving for, something approaching a multiplex understanding, both of the variety of incompatible *fabular* strands and of the precise literal engine (the "gap" at Empire Star) that permits their synthesis into a realistic *sjuzhet*. With that understanding, Jo hands San Severina (back) the red comb.

Illustration: From Paradox to Paratext

To begin to conclude, or at least to anticipate a segue into the following two chapters, I wish to note that all of our descriptions of the multiplexity of narrative reading, particularly when multiplied *sjuzhet* constructions are explicitly irreducible to "historical" *fabulas*, adopt a fundamentally visual metaphorics. This is a key facet of reading easily missed with respect to classical narratives but perhaps impossible to miss in time travel paradox stories. As the referential pathos compels us to "reconstitute," in Genette's term, an ultimately unreconstitutable *fabula*, we end up occupying a position so diagrammatic or maplike in its multidimensionality that the usual metaphors of reading position—perspective, point of view, and so on—become virtually literal themselves. The more the *sjuzhet* calls attention to itself as a tendentious reconstruction of *fabula*—and the more it resists, therefore, the repostulation of *fabular* apriority, a tacit reduction to an os-tensibly historical chain of events—the more it approximates the level of *text* and starts to appear materially. In *Empire Star*, Delany's constant re-minders to the reader that a piece of literature, a book, is being constructed also signify that the (paradoxical) *sjuzhet* itself is tending toward textuality, even as the in-principle unreconstructability of the *fabula* reminds that same reader that he or she occupies a material body, literally holding and viewing a book put together in a more or less arbitrary order. More to the point, then, the *fabula*, as it necessarily becomes *sjuzhet*—or as its postula-tion unravels—also becomes, as it were asymptotically, *paratext*, the physical embodiment and frame of the *sjuzhet*, literally viewed rather than merely read. The reading rematerializes, becomes a multiplex charting of narrative strands: the red comb in chapter 4, the red comb in chapter 11, the young and old San Severinas on pages 77 and 19, and so on. Ultimately we end up with a viewpoint increasingly verging on a literal *point* in space from which one *views* the book itself, arranged in such-and-such a sequence of sen-tences, paragraphs, chapters, and pages. Against the pathos of *fabular* apriority, subverted by paradox yet still demanding reconstruction in a completed reading, the *sjuzhet* draws the text toward its own last "thresh-old,"[62] the ever more concrete medium of its re-presentation, a visual ma-trix or array of its points, lines, doublings, and retracings, a thing *seen* and *held*. On the book's last page, Jewel's final words read: "It's a beginning. It's an end. I leave to you the problem of ordering your perceptions and making the journey from one to the other"—followed immediately by the paratextual mark of the author: "—*New York City, Sept 1965.*"[63]

It is particularly interesting, therefore, to consider the paratextual apparatus of *Empire Star* itself—which is to say, the published book—and the complex or multiplex manner in which it occupies the threshold of narrative representation. One convenient way to do this is to consider different printed versions of the novel. I turn to an illustrated edition of *Empire Star* that appeared in a collection of Delany's work entitled *Distant Stars*, published by Bantam in 1981.[64]

The illustrator of the *Distant Stars* edition of *Empire Star*, John Jude Palencar, renders his selected scenes into quarter-circles that are then, at the end of each chapter, recombined into complete circles. This technique pursues some of the metatextual hints offered by Delany's narrator, opting for moments in which simplex partial views are perceivable as pieces of larger, more synthetic pictures of the story's progress. Sometimes a specific illustration is repeated on different pages, corresponding to instances in which the *sjuzhet* repeats the "same" event in the *fabula*, only viewed in (at least) two different ways. At other times, events in the quadripartite chapter-ending pictures don't appear in prior quarter-circles at all, which creates new perspectives. The overall effect is to reduplicate the complications of the paradoxical plot, in which no single *fabula*—represented here literally by a sequence of pictures—is recoverable, but in which a chain of images on the level of the *text* nonetheless collects a series of multiplex views upon possible alternatives in the *sjuzhet*. In this sense, the illustrator is a good and careful reader of the novel's structural vicissitudes. The circles collectively represent some of the viewpoints-over-histories presumably needed to account for both the difference *and* the identity of incompatible moments in a postulated but chronologically unreconstitutable *fabula*—they are possible mappings of *Empire Star*'s coherence or relative incoherence as a "world."

At the start of the final chapter, the one in which Jewel sets out to recap (for "the multiplex reader") some of the plot's seemingly incompatible directions, the illustrator accomplishes a kind of master stroke, at least in narratological terms. He now reinserts all the previous chapter-ending circles in succession, captioned with lists that specify the viewpoints shown

Empire Star, quarter-circle illustration

by each of the quarter-circles within. Thus spatialized and catalogued, the discrete moments of *sjuzhet* are condensed into something approximating a tableau of the novel's entire history: each linear strand, however irrecoverable on the level of *fabula*, is made the metonymy of the others; therefore a sense of closure, along with a certain arbitrariness in the ordering of both the individual quarter circles and the completed circles, is made explicit. The arbitrariness is aesthetic, of course: the pictures are to be looked at distinctly and timelessly; they do not add up to a sequential strip, as in a graphic novel, but remain persistently listlike. Yet the arbitrariness is also fundamentally structural, a paratextual metastory about the novel's own

Jewel

Jewel and Comet Jo see San Severina
Jewel and Comet Jo meet young San Severina
Jewel, Ki and Marbika in Organiform Cruiser
Jewel becoming crystallized Tritovian

Empire Star, chapter-ending illustrations

narratological complication, and its particular method of utilizing or theorizing the postulate of *fabular* apriority. One sees, in essence, that although
no coherent history is recoverable on the level of *fabula*, a certain history
nonetheless *does* get recovered on the level of text or, more properly, paratext.

San Severina

San Severina observes Comet Jo after his haircut
San Severina says goodbye to Comet Jo
San Severina and seven Lll
San Severina in dungeon is seen by Comet Jo

One has, in the end, a finished novel, a *book*, that stands in as the recourse of "reconstitution" that the *fabula* could not provide, becoming essentially the substitute repository for the referential pathos, or maybe its determinate negation. Thus, although the naturalization of *fabular* apriority is rendered impossible by the time travel paradox story, the story does *not* end in a state of chaos or indeterminacy but rather with a fully concrete paratext, a book, which is a medium (in a rather literal sense) through which one may continue to play the dialectical game of reconciling the temporal orders of *sjuzhet* and *fabula*. And thus, strangely enough, the paratext is even more primal than the *fabula*—or stands in behind it, the ultimate postulated object of the cathexis of narrative coherence or truth, the *thing* one must finally always possess in order to read, and to finish reading.

Indeed, ultimately the conservative or preservative impulse of reading, the desire or need to make a single *picture* out of the story's history, despite illogic or incompatibility, trumps the paradox. The power possessed by the text—or again, more properly, the power possessed by the paratext, that concatenation of physical means through which *sjuzhet* is collated and presented to the reader as actual, usable material—to determine and drive our construction of a singular viewpoint-over-histories may be seen in one more, strange example from the publishing history of *Empire Star*. In 1983, Bantam republished the *Distant Stars* version of *Empire Star* in mass market paperback format. The reader of this new Bantam edition arrives at the end of the penultimate chapter, only to encounter a list of sentence fragments spanning four pages, beginning like this:

Jewel

Jewel in crystalline form meets Comet Jo
Jewel, Comet Jo and Ron
Jewel, Comet Jo and Lump
Jewel in Comet Jo's eye

Jewel

Jewel and Comet Jo see San Severina
Jewel and Comet Jo meet young San Severina
Jewel, Ki and Marbika in Organiform Cruiser
Jewel becoming crystallized Tritovian

Lump

Muels Aranlyde, a part-Lll writer
Muels escapes with Ni Ty Lee
Muels is sold by Ni Ty Lee
Lusp and Ni Ty Lee meet young San Severina[65]

As we can see—but the reader of the 1983 Bantam edition alone could not possibly have seen—these are in fact the captions from the previously illustrated *Distant Stars* edition of *Empire Star*. An editor somewhere in the publication process has removed the illustrations themselves but through some grievous copy-editing mistake has left the captions in. They now appear like an interpolated catalog of fragments, never written by Delany and therefore not part either of the original novel or of its revisions, but taking up their place in his work nonetheless, seeming to recapitulate the narrative's "kernel" moments. In short, they look like part of the *sjuzhet*, some strategic component of the narrator's (Jewel's) attempt to diagram the full set of *fabular* strands for a multiplex reader.

In a sense, this mistake, which is a wholly and egregiously paratextual intrusion into the *sjuzhet*, shows more clearly than anything else—even more than the missing illustrations themselves would have shown—how a multiplex synthesis, a picture or view of the novel's fraught misrelation to *fabula*, might be constructed or assimilated by a reader. Concretizing the transtemporal juxtaposition of otherwise incompatible simplex lines of events, this pseudosection of the novel depicts, quite literally in the form of a spatialized list, at least one concrete version of the viewpoint-over-histories required to narrate or read the plot. For instance, to return to our example of comb exchanges, we can here read in spatial juxtaposition: "Young San Severina is given comb by Comet Jo," "Comet Jo with comb on Rhys," and "Comet Jo gives comb to young San Severina."[66] Quite right: the narrated moments of the *sjuzhet* become text, and then paratext, something literally viewable in a spatial array or matrix, and therefore ultimately an entirely synthetic reconstitution of the novel's otherwise unreconstitutable *fabula*. That the pictures to which these captions refer are missing makes no essential difference to either the sequence or its postulated coherence; the captions function as a "mapping" of the story quite effectively on their own. And they do so precisely because the referential pathos of reading wants *reference* more than it wants either logic or masterful fictionality. The text itself, to pick up James's phrase, "acts like a historian"—acts to reduce plot events to *fabular* apriority—even in a case where a purely accidental paratext intrudes and illegitimately apes the role of *sjuzhet*.

An editor's accident that permanently contaminates an edition of his novel is no doubt a source of irritation or chagrin to Delany himself,[67] as it might be to any reader or scholar of Delany's book justifiably concerned to acquire a full or correct version of the novel. Nonetheless, for the narrative theorist the accident is felicitous, supplying a moment in which an outsider's viewpoint—that of the illustrator, the captioner, the editor, or some untraceable combination of these agents (so much the better that it is unclear precisely *who*)—is literalized within the text as a series of entirely paratextual assertions that become part of the *sjuzhet*. Call this an ultimate instance of time travel paradox fiction, all the more so for its anonymous and illegitimate status: the artificial, synthetic multiversal viewpoint-over-histories, entirely outside of the *sjuzhet*, reinjecting itself into the *sjuzhet* as a literal re-viewing of the chain of events that were primally postulated as forming, and now again will be assumed to form, the *fabula*.

The Primacy of the Visual in Time Travel Narrative

> I told her she didn't exist, and that her whole world didn't exist, it was
> all created by the fact that I was watching, and would disappear back
> into the sea of unreality as soon as I stopped looking.
>
> —GEOFFREY A. LANDIS, "Ripples in the Dirac Sea"

> Reality is that which when you stop believing in it, it doesn't go away.
>
> —PHILIP K. DICK, *Valis*

In the preceding two chapters I have been assessing the merits of my initial proposition that time travel stories are a "narratological laboratory" in which structuring conditions of storytelling are depicted as literal plot. I would like briefly to sum up what one observes in such a laboratory, particularly with respect to my reading of Delany's *Empire Star*. Here are some findings:

First, as the story of *Empire Star* proceeds, the viewpoint of its protagonist, Comet Jo, evolves from a "simplex" into a "complex" and finally into a "multiplex" perspective, with a capacity to roam over and reconcile multiple, even contradictory, lines of the *sjuzhet*. Such an extrasubjective viewpoint-over-histories is structurally required by any narrative that exhibits at least a complex relation between *sjuzhet* and *fabula* (which is to say, virtually any narrative at all), but the time travel paradox story adds the following twist: the viewpoint-over-histories is represented literally, then possibly or actually subverted within the plot, via the perspectives of actual characters such as Comet Jo and Jewel.

Second, as Jo's viewpoint-over-histories develops, it increasingly approaches the position of the novel's reader, whose synthetic perspective also necessarily resides outside simplex lines of events. In Chapter 3, borrowing from time travel authors who invent ways to depict the scenography of transtemporal or transworld travel, I described such a reader's viewpoint as a hyperspacetime position. The initial proxy for such a position in *Empire Star* is the multiplex narrator Jewel, who is freed from the limitations of chronological or causal order, and whose readerly or "literary" reconciliation of *sjuzhet* and *fabula* Comet Jo also comes eventually to approximate.

Third, as Jo's approximation to Jewel's multiplex position develops, the *sjuzhet*, the novel's strategic arrangement of the multiple incompatible strands of *fabula*, verges on text or even paratext, something increasingly like a material illustration of plot events and lines.[1] Although beginning by narrating, ultimately what Jewel does is closer to *depicting* the *fabula* for us, representing it more or less spatially or diagrammatically, even referencing external material features such as page numbers and chapters—hence, also, Jewel's continual references to its own literary endeavors, and even to the published book that it or Jo will eventually produce. Ultimately, the viewpoint-over-histories thrust upon the reader, and represented by proxy through Jewel, evolves into something like that of an actual person holding a book and looking at it: a "view" from a "point" in real spacetime.

Fourth, in light of this hypertextual or paratextual realism, the general notion of viewpoint, which we are accustomed to accepting within narrative theory only as a relatively vague, perhaps suspiciously ocularcentric metaphor, ought carefully to be reconsidered in its literal denotation. In the spirit of the strictly "physical" understanding of narratological conditions I adapted from Einstein's critique of geometry in Chapter 3, I would say this: to read the *sjuzhet* of a time travel story is finally tantamount to viewing the story's paratext—or in other words, the layer of the *sjuzhet* asymptotically becomes this more directly material stratum of narrative organization. The pathos of a postulated *fabula* still drives our reading—time travel is, after all, still a play on classical narrative, and shares the latter's usual compulsion toward coherency and self-conservation. But as the in-principle irreconcilability of *sjuzhet* and *fabula* becomes all too apparent in the time travel paradox narrative, a recourse to the postulate of *fabular* apriority cannot continue to be persuasive. The coherency of the narrative, instead, tends toward a relegation to the mere text—an artificial stitching together of ineluctably self-contradictory *fabular* strands—and therefore eventually to the paratext, which we may now postulate as the true *a priori* medium of conservation.

Fifth, to the degree to which *fabula* ceases to operate as an effective recourse for narrative synthesis, the paratext becomes increasingly *visible*. *Fabular* apriority becomes paratextual apriority, and the reading of *sjuzhet* then becomes explicitly the physical action of leafing successively through the pages (and/or images) of an actual book in spacetime. Now the novel becomes a visual *object*, an illustration of itself in a real location, like our illustrated edition of *Empire Star*, or even like our projection of its diagrammatic structure after the illustrations have been removed. Not coincidentally, the pages, chapters, and generally the physical arrangement of "lines" can all enter the plot directly and be unproblematically referenced by Jewel or Comet Jo, just as they are (sometimes) by the illustrator: diagrammatic depiction has become the story's primary mode. We might then be warranted in speaking of the referential pathos as a *paratextual pathos*, a "return" to the visual paratext that was, after all, the elemental physical medium of the *sjuzhet*'s representation of *fabula*, the actual place where it was first postulated.

This last point is fundamental. The time travel paradox story exposes the postulate of *fabular* apriority—the Jamesian pretense of an underlying and prior historical or factual story—not merely as a fiction of its own, but more crucially as a dissimulation of the real, physical means of prioritizing *fabula*, the paratext. What the time travel story then reveals is also a prior, fundamentally ideological prejudice within "classical" reading, the leading tendency of narratives to postulate metaphysical rather than physical strata of reconciliation, to direct us toward their *fabulas* rather than toward their texts and paratexts, just as realistic painting or photography directs us toward an image rather than toward oil pigments or emulsions. Time travel paradox, disallowing any straightforward postulation of *fabular* apriority but nonetheless straightforwardly maintaining the classical referential pathos, both depicts and subverts this metaphysical bias of narrative reading generally, the latter's preference for postulated story events over physical media, even within the dialectical revisions of metanarration. Simultaneously, time travel paradox exposes the actual physical conditions required to maintain such an illusion (or its revision) in the first place. The paradox story begins by mapping the structural multitemporalities of the act of reading in its hyperspacetime scenography, and ends by showing us the real spacetime of reading in the form of a physical encounter, a book seen by actual eyes, and in it, a literal illustration of its own rearrangement of (postulated) *fabula*.

In sum, the reader's "viewpoint" is therefore no mere metaphor, but is rather something approaching a strict description of the physical conditions of the reading act, a "point" in real spacetime from which books are "viewed" as diagrammatic layouts of their own textual arrangements of *sjuzhet* and *fabula*. Philip K. Dick, whose suspicious, even paranoid, perspective on the relationship between readers and signs generates insightful theory about the transition from text to book, describes such a physics of reading in characteristically hallucinogenic terms: as the observation of a "hologram" in which the "different ages" of a narrative are laid out, simultaneously "along the time axis" and in terms of depth, "in many successive layers"; at the same time, for Dick, the world of the reader becomes like a bound book, the "accretional layers laminated together."[2] In its consummation, which Dick (in his later novels) does not hesitate to extrapolate, such a god's-eye view achieves the *a priori* visualization of *any* possible "layout" of events, along with the insight that the visual assimilation of the whole precedes the "successive" laying out of parts and presumes a "harmonious fitting-together" as inevitable as the "*kosmos* of Pythagoras."[3]

Returning to *Empire Star*, I would propose the following empirically counterintuitive but structurally sound claim: the illustrations of Delany's novel are not added to the text but rather precede it. Illustration is, in fact, the elemental structural medium through which the *sjuzhet* of *Empire Star* can appear to return to its *fabula*, or through which the plot establishes and conserves itself in the form of a finished novel. No wonder, then, that the mistakenly leftover captions, those apparently accidental paratextual appendages, can still operate as—or, if we were to follow Dick's lead, can allude back to—a coherent reading of the novel's plot. The pictures of events to which these leftover captions refer are structurally prior to the *fabula* they set out to illustrate; in essence, *Empire Star postulates* them. In this light, I suggest the following specific reversal of James's claim about the historicist vocation of the novelist, or at least a reversal of what that vocation yields for the reader: first comes a "picture" of the *fabula*, in the guise of a paratext held and viewed by a physical reader in space and time, with pages, chapters, and a layout; *then* comes the *sjuzhet* presented "within" a text and "viewed" from a readerly position in hyperspacetime (represented in the time travel story by a proxy transtemporal traveler); last comes the narrative's *fabula*, always postulated as prior, even when its "historical" disposition is rendered permanently indefinite by paradox. Thus time travel adds only this distinct but crucial aspect to our understanding of classical narratives: with the postulate of *fabular* apriority indefinitely deferred, it is the (visual) paratext that stands in for the primal scenography of narrative; if a paradox story obstructs the "natural" tendency of the referential pathos, then paratext is the layer to which the narrative ultimately returns.

Therefore, when I now shift my attention to visual depictions of time travel in television and film, these media should by no means be construed as late or supplementary modes of time travel fiction; they are rather its very ground, in structural-narratological terms. Critics have often suggested, as Brooks Landon writes, that "cinema itself is a kind of time machine" or that "the primary special effect of film is always one of time travel or time manipulation."[4] My arguments extend Landon's observation but also partly reverse it. Time travel, even in the form of unillustrated text, is already fundamentally a visual medium, a literal depiction of the textual and paratextual conditions under which viewpoint is constructed. "Reading" its plot is already like "seeing" its illustration, its diagram, or its film in physical spacetime. Thus what we shall observe, as we further explore the laboratory of the time travel story in visual formats, is the construction, deconstruction, and reconstitution of viewpoint that even the most mainstream time travel narratives ineluctably theorize as they offer themselves straightforwardly for a viewer's perusal and pleasure.

FIVE

Viewpoint-Over-Histories: Narrative Conservation in *Star Trek*

My chief examples in this chapter are taken from the television series *Star Trek: The Next Generation*, which ran from 1987 to 1994 and featured time travel plots in a number of episodes. I also discuss some ways in which analytic philosophers have treated time travel stories as thought experiments in logic and causality, and some parallel theorizations within both popular fictions and physics. The eventual convergence between analytic philosophy and, for instance, a popular television series will not be far-fetched. Both philosophy and mainstream television treat the potential for paradox as a means of demonstrating what passes for acceptable narrative progression and conclusion, and both are profoundly concerned with preserving the logical propriety of such narratives, their chronological and narratological coherence. In this sense, the most complex analytic philosophy and the most straightforwardly generic television or popular film share an abiding concern with the avoidance of paradox even as they indulge in it, a curious ambivalence that often emerges in the form of paraliptical pseudoclaims:

here is a plot event that is logically impossible; here is a worldline that could never have existed; here is an alternate history that is "fiction," and its (fictional) counterpart is "fact."

The best-known version of a time travel paradox is the "grandfather paradox," in which an individual goes back in time and kills his or her own ancestor, making it impossible for the traveler him- or herself ever to have been conceived. In its general layout, this is a paradox of causality: the time traveler causes a scenario, the effect of which is to eliminate the cause of that very scenario—namely, him- or herself. Here is a pithy version by Miriam Allen deFord: "No, you can't go back into the past and kill your grandfather, as people used to fancy, for the very good reason that if he had been killed, you wouldn't be alive to make the trip. You would never have been born."[1] And here is another version, characteristically more elaborate, from an editorial preface by Hugo Gernsback in 1929:

> Suppose I can travel back into time, let me say 200 years; and I visit the homestead of my great great great grandfather, and am able to take part in the life of his time. I am thus enabled to shoot him, while he is still a young man and as yet unmarried. From this it will be noted that I have prevented my own birth; because the line of propagation would have ceased right there.[2]

It seems obvious, yet is rarely pointed out, that the presence of grandfathers in such examples instead of fathers or mothers is an odd dissimulation of the Oedipus complex.[3] Killing one's parent would certainly be a more reliable way to ensure paradox than eliminating a grandfather, let alone a "great great great grandfather," since one could sidestep the potential errors introduced by the overpopulation of the narrative with unconfirmed or tenacious predecessors.[4] Paul Horwich's "autoinfanticide" and Peter Vranas's "retrosuicide" are attempts at more efficient paradox, in which one travels back and kills one's own earlier self, even if such cases might present fiction writers with further difficulties, in terms of motive, than parricide.[5] But avoiding such questions of motive may be precisely the point for both fiction writers and philosophers. A grandfather, like a younger self, represents a progenitor sufficiently uncathected that paradox may be constructed less as a psychological than strictly as a logical, or perhaps narratological, event. Oedipal romance is thus sacrificed for, or submerged within, logical or narratological enigma, much in the same way that early time travel writers sacrificed or submerged the sociopolitical content of late utopian romance in favor of the macrological exposition of narrative framing.

From early on, time travel paradoxes have nearly always been set up as conundrums of sequence or order, and far less often as strictly psychological, let alone erotic, crises. This has tended to be true even where oedipal eroticism is directly mentioned or parodied, as I will discuss in my reading of *Back to the Future* in the next chapter. This is not to suggest that either science fiction authors or physicists and philosophers are unaware of the oedipal (or narcissistic) grounds of their plots or speculations, but rather that such grounds tend to become manifest, when at all, in the form of glib jokes or twists, droll asides to the serious *logical* complications that are more likely to be foregrounded. In essence, science fiction is highly adept at—arguably, it may be founded upon—the suppression or sublimation of its own primal scenes, or their transmogrification into plot puzzles. Any adequate critical or theoretical interpretation of time travel paradox stories will therefore generally be obliged to dig out and reconstruct the more primal motives or grounds of these stories.

I can begin such a critical or theoretical reconstruction by noting the basic conservatism built into most time travel narratives, a conservatism to which I also alluded in the prior chapter's reading of the postulate of *fabular* apriority. Within time travel paradox stories, the historical past tends to preserve or to protect itself, even to "heal" itself when necessary, but in any case to persist as either what it *is* or what it was *supposed to have been,* according to a range of narrative, moral, or aesthetic logics that will have to be analyzed in turn. Likewise, and presumably not coincidentally, pre-oedipal order, along with its corollaries of unidirectional genealogy and decathected historical consistency, is also confirmed or reestablished, a tendency that colludes with the conservative stylistic or generic leanings of popular literature more generally.

In short, if characters or events in a time travel story conspire to change the past, very often other characters or events will intercede to forestall such a change or to counteract its influence: paradox will be repaired. A time-traveling protagonist in Poul Anderson's *There Will Be Time* declares, "I've tried altering the known past and something always happens to stop me"; elsewhere Anderson writes that "the course of the world has enormous inertia."[6] A character in John Varley's "Air Raid" suggests that "we can do

things in the past only at times and in places where it won't make any difference"; "events are conserved," declares a Damon Knight character in "Arachron"; and in Fritz Leiber's *The Big Time*, history is subject to a "law of the
Conservation of Reality."[7] Nearly as often, authors provide macrological
explanations or even quasi philosophies of such conservation: in "Vintage
Season," C. L. Moore writes that "the physiotemporal course tends to slide
back to its norm."[8] Sometimes time or history themselves are the agents of
self-healing: Bill Pronzini writes that time itself can "seal the apparent rent
in its fabric" and even "unmurder" a paradoxically dead ancestor; Michael
Moorcock suggests that "if one goes back to an age where one does not
belong, then so many paradoxes are created that the age merely spits out
the intruder as a man might spit out a pomegranate pip which has lodged in
his throat."[9] F. M. Busby provides this colorful allegory: "The past—it's
pretty damned solid. . . . It's a little like a compost pile—fairly soft near the
surface, but packed hard further down, with all that Time piled on top of
it."[10] John Brunner speaks of alterations of the past as "sluggish events,"
L. Sprague de Camp of events "too well rooted to be destroyed by accident,"
and Connie Willis of a "tough, immutable past."[11] As the incongruous
metaphors suggest, the inherent ambivalence of the "natural" situation that
these authors depict—that altering the past *can* be done by the time traveler
in theory, but nonetheless *wasn't*—lends itself to hyperbole and caricature.
Even small disruptions of the spacetime continuum are often rejoined by
absolute catastrophe, as in Larry Niven's "Rotating Cylinders and the Possibility of Global Causality Violation," in which a character who attempts
to create a paradox is removed by a coincidental nova; or in Fredric Brown's
"Experiment," in which a scientist tries to engage a bilking paradox with a
small time-traveling cube, and the "entire rest of the universe" (except for
the cube) vanishes.[12]

Whether subtle or extreme, such conserving maneuvers, designed either
to restore the past or to protect it from potentially illogical interference,
are altogether common in the literature. Presumably not coincidentally, *dei
ex machina* are perhaps more readily tolerated in the time travel paradox
story than in just about any other twentieth- or twenty-first-century literary form. On the one hand, this conservative tendency might seem surprising, given the considerable opportunities for the manipulation of sequence,
causality, and historical precedence that time travel fiction appears to open

up. On the other hand, conservation might seem like an inevitable compensation, through which time travel fiction pushes back against its own drive to become more radical, subversive, or chaotic than its customary niches in popular fiction might tolerate. Either way, the conservatism of time travel paradox stories seems to indicate a peculiar crisis of the *present* within their narrative logics, a crisis in which generic expectations collide with a heightened drive toward narratological radicalism. If time travel is possible at all in a story, then in principle anything could have happened in the past other than what *did* happen; it is therefore all the more urgent that an "actual" singular past be restored to its natural *fabular* status—"natural" because it is the past of *this* present. "Nature despise[s] a paradox," as Dean Koontz suggests, although the apothegm is finally more applicable to the generic nature of *narratives* than it is to nature per se.[13] Of course, given that time travel stories open up the possibility of altering or damaging this "natural" state, even of literally rendering it *ex post facto* impossible, they also offer the opportunity to critique the conditions (narratological, psychological, and ideological) under which stories find themselves compelled to repostulate it. I wish to observe precisely to what degree and, more crucially, by what means and by virtue of what unnatural narratological lengths the "natural mechanism"[14] of stories is maintained, once paradox enters the picture.

Conserving Fabula *in "Time's Arrow"*

The double episode of *Star Trek: The Next Generation* entitled "Time's Arrow" (1992) begins with a discovery opening the possibility of paradox but then works hard to shut it down, or, more precisely, works to indicate how and why no paradox could ever have occurred.[15] In the twenty-fourth century (the show's present moment), the android Commander Data's head is discovered lying in an abandoned mine in San Francisco, seemingly untouched since sometime late in the nineteenth century. The simultaneous presence onscreen of the abiding Commander Data (intact, of course) and his 500-year-old vestigial head may already indicate a paradox of sorts: Data's head is in two places at the same time, both on and off his neck, and the early scenes are constructed to take advantage of this uncanny coincidence. But this isn't yet what one calls a time travel paradox per se. Indeed, in terms of time travel, the spatial simultaneity of the two temporally divergent heads is resolvable (by Data himself, as it turns out) through a narrative maneuver:

> Data: It seems clear that my life is to end in the late nineteenth century.

> Riker: Not if we can help it.

> Data: There is no way anyone can prevent it, sir. At some future date, I will be transported back to nineteenth-century Earth, where I will die. It has occurred; it will occur.

Data's assertion "it has occurred; it will occur"—or to combine the tenses into their implied future perfect, "it will have occurred"—captures precisely the inevitability or facticity, the pastness, of the historical events confirmed by the incontrovertible fact that Data's 500-year-old head has lain untouched in that cavern. From here, the episode's *sjuzhet* sets out to explain what brought Data to nineteenth-century San Francisco and, as it turns out, what then returned his head five centuries into the future to be reunited with his decapitated body. Ultimately, there will have been only one head, one past, and one *fabula*, all restored to coherent linear sequencing via looped interventions that, although they may have affected the past, in the end will not have changed it.[16] More precisely, the effects engendered by the loop will have affected the past just enough to make it what it had always been. In the episode's final scene, a head still lies abandoned in a nineteenth-century mine, and 500 years later the same head will have been reattached to Data's neck, so that both past and present end up exactly what they were when the episode began. The postulate of *fabular* apriority is thus reconfirmed, albeit uncomfortably (for a postulate) after the fact, through the roundabout closure of the *sjuzhet* of the episode itself.

This sort of plot, in which characters or actions affect the past but do not change it, is a favorite of both physicists and philosophers because it appears to be consistent both with logic and with the theoretical possibility of "closed timelike curves," solutions of the general theory of relativity that permit time travel to the past. A closed timelike curve is a path along which one may effectively travel faster than light, whether through a wormhole or by some other exotic technology, and therefore return to a spacetime earlier than the one from which one departed. The theoretical possibility of closed timelike curves has been accepted, with varying degrees of enthusiasm, since Kurt Gödel's proposed solution to general relativity in 1949, but has experienced an increased theoretical cachet following the work of a group of physicists from the mid-1970s to the 1990s, especially Stephen Hawking, Igor Novikov, and Kip Thorne. *Star Trek*'s "Time's Arrow" is both cognizant and respectful of such physical theory, offering a time travel loop in which causal order is not upset, or, in other words, in which no strictly logical paradoxes ensue. Even though potential paradoxical branchings of the *sjuzhet* frequently suggest themselves—Data could easily decide, for instance, not to go back to the nineteenth century, or Picard or Riker could step in and prevent him from doing so—each alternate remains a mere counterfactual hypothesis. Ultimately there is only one unchanged past postulated by the episode's present.

The philosopher David Lewis articulates what has become the canonical formulation of such a singular, logically consistent model of time travel: "Could a time traveler change the past? It seems not: the events of a past moment could no more change than numbers could."[17] Lewis elaborates with the following anecdote, which, like *Star Trek*'s "Time's Arrow," at first appears to leave open the hypothetical possibility of paradox:

> Consider Tim. He detests his grandfather. . . . Tim would like nothing so much as to kill Grandfather, but alas he is too late. Grandfather died in his bed in 1957, while Tim was a young boy. But when Tim has built his time machine and traveled to 1920, suddenly he realizes that he is not too late after all.[18]

Quite typically (even though this is philosophy and not popular fiction), the time travel tale initiates a crisis in the apriority of the *fabula*, which now as it were invades the *sjuzhet* too directly and is thus subject to the sort of free manipulations of order that *sjuzhet* (but not *fabula*) allows. Tim has every opportunity—at least it appears so—to ruin the "natural" order of what had started out as the history of this plot, namely his direct descendance, in his own *fabula*, from the detested grandfather:

> Tim can kill Grandfather. He has what it takes. Conditions are perfect in every way: the best rifle money could buy, Grandfather an easy target only twenty yards away. . . . What's to stop him? The forces of logic will not stay his hand! No powerful chaperone stands by to defend the past from interference. (To imagine such a chaperone, as some authors do, is a boring evasion, not needed

to make Tim's story consistent.) . . . By any ordinary standards of ability, Tim can kill Grandfather.[19]

However, as Lewis explains, the word "can" is equivocal: Tim, with his time machine and a perfectly ordinary set of skills, certainly has the "ability" to kill his grandfather, but simultaneously not the "possibility" of doing so. This equivocal "can" operates in something like the sense in which Lewis *can* speak Finnish (in terms of "facts about my larynx and nervous system") but nonetheless *cannot* speak it (in terms of "my lack of training").[20] Thus, "Tim's killing of his grandfather that day in 1921 is compossible with a fairly rich set of facts"—Tim's having successfully arrived in 1921 via the time machine, his skill as a marksman, and his grandfather's presence in the gun sights.[21] But it is "not compossible with another, more inclusive set of facts"—"the simple fact that Grandfather was not killed," "Grandfather's doings after 1921 and their effects," and so on.[22] Hence no paradox, because no (past) murder:

> Tim cannot kill Grandfather. Grandfather lived, so to kill him would be to change the past. But the events of a past moment are not divisible into temporal parts and therefore cannot change. Either the events of 1921 timelessly do include Tim's killing of Grandfather, or else they timelessly don't.[23]

Since the past is by definition "timelessly" what it was and (therefore) is— "the one and only 1921"—changing it would involve a straightforward logical contradiction, a 1921 in which the grandfather was both killed and not killed (A and not-A). Thus it is "logically impossible that Tim should change the past by killing Grandfather in 1921," even if the effects of Tim's intervention are themselves part of that singular historical record.[24] "He isn't going to do it because he didn't," as Murray Leinster says of a similar temporal intervention: "He can't because he hasn't."[25] A Robert Heinlein protagonist concurs: "I *wouldn't* because I *didn't*," and a James Blish character says, of a certain vital secret, "you didn't find it for one reason, and one reason alone; because you didn't find it."[26] Very often, characters discover that they themselves had been the unwitting cause of the very phenomenon they now observe in the present or future, like the Leo Frankowski character who discovers that "I was responsible for introducing the polka into Poland,"[27] or more poignantly, the protagonist of Chris Marker's film

La jetée, whose childhood memory of witnessing an assassination compels him eventually to return to the very time and place of the shooting, where he discovers that he himself is the victim.[28] Just so, Data's head, having been tossed to its particular spacetime spot in the cavern, will always and forever have been lying there from the nineteenth century to the twenty-fourth; the interventions of the time-traveling *Enterprise* crew are already constituents of that ineradicable historical record, and will add nothing further to it. Never mind whether one "can" tell a different story or hypothesize an acausal intervention—that is mere supplemental *sjuzhet*, mere counterfactual speculation.

For science fiction writers and philosophers, the authority of the model articulated by Lewis, which Peter Vranas appropriately identifies as the "standard solution" of time travel paradoxes,[29] is considerably enhanced by the support of physicists following Igor Novikov, who in 1979 argues that "the close of time curves does not necessarily imply a violation of causality, since the events along such a closed line may be all 'self-adjusted'—they all affect one another through the closed cycle and follow one another in a self-consistent way."[30] By 1998 Novikov is able to write that his own self-consistency principle "appears to have been accepted by everyone who works in the time machine field," and that "I and my colleagues were able to prove that this principle can be deduced from the fundamental laws of physics."[31] However, the variety of views on "chronology protection" in this "time machine field" makes Novikov's sanguine statement about the general approval of his principle slightly contentious. Another group of physicists prefers the related but not identical view that chronology is protected from inconsistency by the very impossibility of time travel itself. Thus in 1991 Hawking proposes a "Chronology Protection Conjecture," arguing that "quantum effects are likely to prevent time travel" or, more strongly, that "the laws of physics prevent the appearance of closed timelike curves" altogether.[32] Hawking asserts, "It seems there is a chronology protection agency, which prevents the appearance of closed timelike curves and so makes the universe safe for historians"; and, again with his tongue characteristically in cheek, "There is also strong experimental evidence in favor of the conjecture from the fact that we have not been invaded by hordes of tourists from the future."[33] Kip Thorne is more diffident than Hawking, suggesting in 1994 that "the laws of quantum gravity are hiding from us the answer to whether wormholes can be converted successfully into time machines."[34] Finally, other physicists such as David Deutsch propose theories, indebted to quantum-theoretical models such as Everett's "relative state" hypothesis and Feynman's "sum over histories," which assert that "for time travel to be physically possible it is necessary for there to be a multiverse."[35] Incompatible causal lines would then be relegated to parallel or branching universes in which they could no longer logically interfere with one another, and unimpeded *fabular* order would be restored for *this* universe.

What logicians and scientists do generally agree upon is that straightforward causal paradoxes such as the grandfather paradox are prohibited by both logic and physical law, whether or not time machines themselves are either theoretically or practically possible. In all cases, the past of a historical line (within a single universe) is sacrosanct, "timelessly" what it was and always will have been. As Richard Gott states in this vignette, the past is comparable to a recorded film:

> Think of rewatching the classic movie *Casablanca*. You know how it's going to turn out. No matter how many times you see it, Ingrid Bergman always gets on that plane. The time traveler's view of a scene would be similar. She might know from studying history how it is going to turn out, but she would be unable to change it.[36]

I will have occasion to reexamine this illuminating analogy between historiography and cinematography in the following chapter. For now I wish to focus attention upon a few of the narratological implications of the logical inevitability of the past, pursuing the Novikov/Thorne/Hawking model that Gott continues to identify as the "conservative approach":

> If [the time traveler] went back in time and booked passage on the *Titanic*, she would not be able to convince the captain that the icebergs were dangerous. Why? Because we already know what happened, and it cannot be changed. If any time travelers were aboard, they certainly failed to get the captain to stop. And the names of those time travelers would have to be on the list of passengers you can read today.[37]

What makes the approach Gott describes "conservative" is both its reliance on a single or "block" universe model of spacetime and its tendency to preserve an already "recorded" past through both physical and logical means.

If time travel threatens past change, then some presumptive mechanism of prevention intervenes, or rather always already will have intervened; therefore, also, within that "recorded" past there must (already) be the traces of the means by which paradox was prevented. Lewis continues his own quasi-science-fictional anecdote in a manner precisely like Gott's *Titanic* story, but in line with Novikov's interventionist model: minor plot events would have to intercede in order to preserve the "consistency" of the major events of the narrative, most importantly the survival of Grandfather:[38]

> You know, of course, roughly how the story of Tim must go on if it is to be consistent: he somehow fails. Since Tim didn't kill Grandfather in the "original" 1921, consistency demands that neither does he kill Grandfather in the "new" 1921. Why not? For some commonplace reason. Perhaps some noise distracts him at the last moment, perhaps he misses despite all his target practice, perhaps his nerve fails, perhaps he even feels a pang of unaccustomed mercy.[39]

Or, such prohibition might occur through something like Hawking's "chronology protection" mechanism, which prevents time travel altogether (Hawking: "Every time machine is likely to self destruct . . . at the moment one tries to activate it");[40] or it might occur by more minor interventions that forestall inconsistency even in light of the time traveler's past intervention (Thorne: "[S]omething has to stay your hand as you try to kill your grandmother");[41] or it might occur by an accommodation of multiple outcomes in a physical multiverse (Gott: "The time traveler just moves to a different universe, where he will participate in a changed history").[42] One way or the other, Tim's story is already "filmed," to use Gott's metaphor; its editing is already complete and fully available, at least in principle, for reviewing: history "is not past; it is always going on . . . you just have to be watching the right screen."[43] Similarly, the presence of Data's head in the nineteenth-century San Francisco cavern is as immutably "recorded" and unchangeable as the television episode itself, fixed in video pixels and fated to re-run exactly the same way for whatever portion of eternity network scheduling and advertising revenue will permit.

What, then, accounts for the lingering impression, so crucial for generating time travel fictions, that paradox is always imminent, that plots constantly teeter on the verge of illogic whenever a time traveler confronts a potentially changeable situation in the past? Why does it even seem arbitrary, despite all logic and physics, to assert categorically that Tim *cannot* shoot (or cannot have shot) Grandfather, or that Picard or Riker *cannot* stop (or cannot have stopped) Data from transporting himself to Earth, or that a character or writer *cannot* otherwise compel a *sjuzhet* to go a different way— indeed, a way that would permanently destroy the logical coherence of its *fabula*? Usually both philosophers and physicists have phrased this problem in terms of free will, or as a matter of the capacities of the time traveler, which in principle ought to be similar to the capacities of anyone else in the same situation.[44] One artifact of this resort to free will or capacity is the lingering impression that the creation of paradox is "merely" a matter of an individual's telling the story differently, and that therefore nothing in particular constrains the narrative from proceeding any way it likes: Tim "has what it takes" to kill his grandfather, and "what's to stop him?"[45] Indeed, nothing is to stop him, or at least nothing that appears within the unfolding *sjuzhet* of the "personal time" of Tim's own version of events. Certainly "the forces of logic will not stay his hand," as Lewis asserts, nor will the mere postulate of *fabular* apriority.[46] We might here rephrase the time travel paradox as one of an incompatibility of individual beliefs and choices, especially those of a multiplex observer who is able to compare and contrast Tim's possible or impossible actions with one another: "The facts . . . do not square with the belief that we *can* choose efficaciously," as William Grey asserts about Lewis's story of Tim.[47] But more than the appearance of free will within particular *sjuzhets*, it is narrative itself, or its requisite construction of a paratextual viewpoint-over-histories, that gives us the impression, logical or not, that Tim can successfully murder his grandfather. For we acquire such an impression through *any* counterfactual juxtaposition of two distinct lines of events, such as the two that Lewis identifies as the "personal time" and the "external time" of Tim's story, each with its different application of the equivocal "can."

Here the force of the postulate of *fabular* apriority—the inertial tendency of the *fabula* to remain historical and therefore logically and causally sacrosanct—runs smack into the quasi-psychological dynamism of narration itself, and the impression that narrators are, after all, just like real persons, with choices and wills of their own. I recall a student of mine, when I was teaching Toni Morrison's novel *Beloved*, asking, with some degree of irritation, why Sethe killed her own child: "Why did she have to do that?" Of course, in terms of the phenomenology of reading, and of the sense of *presence* that the narrative subject achieves within it, the student is perfectly justified in asking such a question: Sethe didn't have to do it, within the "personal time" of her own narrative. On every other register— *fabular*, logical, psychological, sociological, political, ideological—and even in terms of the sheer empirical facticity of the book itself, its publishing, production, and distribution history, by virtue of which the printed novel in the student's hands is a *fait accompli*—of course Sethe *must* do it, or rather, *must have done it*. Paradoxical intervention here, à la Lewis's Tim, would negate everything important or inevitable about Morrison's novel. And yet even here the "historical" inexorability of Sethe's act can still seem arbitrary, an artifact merely of her free will, and of whatever suspense we retain concerning the possibility that she might exercise it differently.

The problem, then, is not primarily logical after all, for in logical terms there is clearly only one 1921, just as (all the more) there is only one history of Sethe's past, which continues to remain (timelessly) what it was. The problem is rather *narrato*-logical, for in the *sjuzhet* of Tim, the rules of storytelling, unlike the rules of logic, do not provide an immediate or obvious means of adjudicating the relative plausibility or consistency of either of the two 1921s. Phenomenologically, the powerful anthropomorphism of narrative viewpoint clearly tends to trump the sanctity of *fabula*, a trend that is itself the basic feature of any potential realism in time travel paradox stories, despite their antirealistic illogic or acausality. In terms of the act of storytelling—which is to say, from the perspective of the *sjuzhet* (Tim's "personal time"), which the reader follows in reconstructing the plot—a logically impossible *fabula* (Tim kills Grandfather) may appear just as consistent as a logically possible or necessary one (Tim fails to kill Grandfather). Only a preference, not to say a prejudice, for logic as a regulator of story construction—concomitant with a classical postulate of *fabular* apriority— can count as an aesthetic motive for excluding the former. Rendered as a science-fictional element within a realistic story, such a regulative logic might be what Wayne Freeze ironically terms a "Theory of Useless Information,"[48] at least from the standpoint of a reader or critic. Ultimately, logical stories are not necessarily more plausible than illogical ones, precisely because our filmlike perspective enables both of them freely and impartially. The factical presence of events in an actual *plot*, which is also to say, in a text and a paratext, trumps their ostensible possibility or impossibility within a history.

Even more crucially, the narrative itself is responsible for creating the impression of possible paradox, because it is the narrative that brings into being the paratextual position of reader or viewer for whom no single version of *fabula* is sacrosanct, and who senses the arbitrariness of any postulation of *fabular* (or historical) apriority. Here the genre audience might possess more functional tools than the logician or the physicist for the purposes of reading or viewing: *both* stories of Tim are fictions, and logical fictions are not necessarily better than illogical ones; fact and counterfact are equally subject to the technology of reading or viewing, such as it may be constructed in any given time travel tale. Logic is just not an especially effective criterion for evaluating either the propriety or the potency of plots, even within "scientific" fiction.

Viewing Fabula *in "The City on the Edge of Forever"*

The viewpoint we necessarily occupy in assimilating the time travel tale thus renders both (ostensibly) factual and counterfactual trains of events narratively feasible. We might then be required to revise Gott's figuration of history as "filmed"; in principle, counterfactual histories are "filmable" with the same generic or narratological credibility as their factual counterparts—as Jean Mitry notes, in the cinema "everything is actual."[49] Any perusal of the ease with which time travel films juxtapose the factual and the counterfactual will serve to confirm this general equivalency, for instance the famous *Star Trek* (the original series) episode "The City on the Edge of Forever," written by Harlan Ellison. Having traveled back to 1930s Earth to chase down Dr. McCoy, who has inadvertently changed the past, Captain Kirk and Mr. Spock discover that they must prevent McCoy from saving the life of Edith Keeler, a highly sympathetic character, in order to restore the proper history of their own future. Most interestingly, Kirk and Spock determine the necessity of this course of action, and pinpoint the crucial moment of historical divergence, by *viewing* the effects of McCoy's intervention in newspapers and newsreels made conveniently accessible through a future-viewing video player that Spock cobbles together, back in

1930, out of his tricorder and some scrounged vacuum tubes. Their survey of alternate history through this remarkable cinematic contraption reveals the chain of events that will ensue if Kirk and Spock fail to prevent McCoy from saving Edith: the success of Germany's atomic bomb program in World War II and the eventual triumph of the Third Reich. However thoroughly miraculous such a viewing machine may be, all of the episode's incompatible plot lines are rendered equally present through the moving images it produces, and which the television audience is also permitted to review. Gott's assertion that we "know how it's going to turn out" and are "unable to change it," whatever its logical or philosophical merits, nonetheless does not account for the intuitive ease with which we are, in fact, able to view what will turn out *not* to have happened, indeed with no more difficulty than we view flashforwards or dream sequences. Both the viewer and the logician perhaps require, then, some more subtle or nebulous criteria—extra-historical data? multiple corroborating reports? further genre expectations?—in order to decide which "film" is the historical one and which the alternate. The logical imperative to select between them, in a "conservative" story, can appear as nothing more than a directorial imposition, a choice required for aesthetic purposes, a *postulation*, more or less suitable to a persuasive or elegant *plot*.

Moreover, it is the reader's or viewer's own presence in the narrative, driven or sustained by the referential pathos, that undergirds the "filmed" equivalence of factual and counterfactual *fabulas*. We are *in* that nineteenth-century cavern, and we see Data's head as Data and Picard see it; we are *in* the room with Spock and Kirk, watching newsreels of the German triumph; such a viewpoint is required for the story, not least for generating the horrible alternatives that ultimately render a conservative conclusion desirable. In essence, we are "the mythical superobserver . . . [who is] on the outside and perceives supertime."[50] But such a viewpoint is also the very same one that introduces the illogic of potential paradox in the first place, the impression of divergent grandfathers or Datas or McCoys. Viewpoint-over-histories is both the relic and the engine of paradox, a literalization of the (at least) doubled perspective that all narratives, structurally, allegorically, or (in the case of time travel fiction) literally, enact. Seeing multiplexity in the *sjuzhet*, I now find myself free, like Lewis's Tim, to see it in the *fabula*, and I therefore also observe the arbitrariness of any postulation required to restore the *fabula*'s singular apriority. Finally, then, what I see is

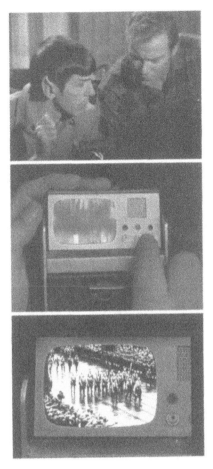

Star Trek, "The City on the Edge of Forever," future-viewing video device

the juxtaposition of the potentially incompatible lines of *fabula* in the text itself, and the wholly material, paratextual means by which they are brought (back) together. I see the filming of the counterfactual *sjuzhet* and the editing of its multiplied lines back into a single story, along with the production and distribution of its paratextual representation in a book or on TV. Paratext, therefore, not *fabula*, is ultimately what is conserved by *sjuzhet*, both because of and regardless of paradox. A film, a television episode, a diagram, a paperback; it could have been otherwise, perhaps it cannot be what it says or shows at all, yet it nonetheless *is*.

Restoring Fabula *in "Yesterday's Enterprise"*

We can examine this paratextual metaseeing and counterseeing more directly if we turn to another episode of *Star Trek: The Next Generation* in which a deliberately paradoxical line of *fabular* events, one that cannot logically exist or have existed in the story's postulated "actual" history, is nevertheless depicted directly on the level of *sjuzhet*—in short, an "illogical" episode. In "Yesterday's Enterprise," the crew encounters a spacetime "rift," and suddenly observes emerging from this rift a ship it recognizes as its own lost precursor, the *Enterprise-C*. That vessel, as history records, was destroyed two decades earlier while coming to the aid of Narendra III, a Klingon outpost under attack by Romulans:

> Data: Sensors confirm design and specifications, Captain. Analysis of hull and engine materials conform to engineering patterns and methods of that time period.
>
> Wesley Crusher: But that cruiser was destroyed with all hands over twenty years ago.
>
> Data: Presumed destroyed. The *Enterprise-C* was last seen near the Klingon outpost Narendra III, exactly twenty-two years, three months, and four days ago.
>
> Riker: And now they're here.
>
> Picard: Has it been adrift for all those years, or has it traveled through time?
>
> Data: It is a possibility, Captain. If that hypothesis is correct, the phenomenon we have just encountered would be a temporal rift in space.
>
> Picard: A rift?
>
> Data: Possibly the formation of a Kerr loop from superstring material. It would require high energy interaction in the vicinity for such a structure to be formed. The rift is certainly not stable, Captain. It could collapse at any time.

In the fictional backstory of the overall *Star Trek: The Next Generation* series, the *Enterprise-C*'s defense of Narendra III had been a pivotal event, precipitating a détente between the Federation and the Klingons and a subsequent period of peaceful coexistence. Now, with the *Enterprise-C* retroactively removed from Narendra III by the rift, a Klingon–Federation war will have ensued and continued twenty-two years into the episode's present moment; that present itself must necessarily be altered. And so it is: a shift between "real" and "alternate" presents is observed directly by the television viewer, who first watches the *Enterprise-C* emerge from the rift, then the present *Enterprise-D* suddenly transmogrify into a war vessel— uniforms, instruments, even lighting instantaneously change—with no knowledge, on the part of the crew, of things ever having been otherwise. We observe a similar special-effect shifting of reality in any number of time travel films and shows, which almost always mark the moment of

Star Trek: The Next Generation, "Yesterday's Enterprise," the temporal rift

transition technologically, for instance with a blurring, swirling, or saturation of the screen image. What is vital, of course, is that we ourselves observe the change from the true history to its alternate, and that we retain a memory of the original, a viewpoint-over-histories entirely unavailable to the characters in the episode (with one exception, as I will discuss). Such a viewpoint, although strictly unphysical, is not only permitted but rigorously demanded by the narration, as I showed already in my reading of "By His Bootstraps" in Chapter 2. Without this metaview, the story could not proceed: there would be only one simplex line to observe, hence no alternate *fabular* possibilities against which to compare it, let alone to which to restore the new, divergent one. Indeed, we would be stuck with the merely immanent and immediate perspective of the crew itself, no better off, in principle, than Thomas Disch's "catatonic who resurrects some part of the past in all its completeness, annihilating the present moment utterly," or Philip K. Dick's "helpless, passive" schizophrenic, to whom "reality happens . . . without relief."[51]

The character Tasha Yar (Denise Crosby), who was supposed to have been dead for several years, suddenly reappears when the *Enterprise-D* undergoes its shift; within this altered history, she was not killed. Thus, among the other abnormal lines of events the episode will be obligated to redress is Tasha's very presence in the show at this moment of *sjuzhet*, not to mention the actor Denise Crosby's presence in the text and paratext. At all levels, her being there is inconsistent with both the story's proper *fabula* and the ongoing history of the *Star Trek: The Next Generation* franchise; this is no longer the "right" storyline, based on everything we know of *Star Trek* and indeed of television itself. The script solves the interesting multi-layered dilemma it has constructed by arranging for Tasha to go back through the rift with the *Enterprise-C* to be rekilled, much as Data's head ends up back where it is supposed to be in the nineeenth-century San Francisco cavern. The Klingon war is (or will have been) averted, the present *Enterprise-D* is restored to (or will have always been) a ship of scientific exploration rather than war, and Tasha Yar is (or will have been) dead and gone. But unlike in "Time's Arrow," here we directly witness both counterfactual divergence and factual restoration as viable plots and image-sets— we view the two *fabular* lines side by side, as it were equivalently. The difference between this plot and Data's (or Tim's) is thus the fully real quality of the alternate, paradoxical universe postulated by the narrative, and its seemingly equivalent capacity to proceed toward a possible conclusion, one in which things might not be as they are supposed to be. This real sense is sufficient to turn "Yesterday's Enterprise" into what certain critics would still call an "illogical" story, but let us note precisely in what narratological terms its difference consists. The transition from a logical to an illogical time travel plot is merely a matter of the degree to which its tendencies toward conservation are represented, not of any qualitative change in its structure. All the same narrative elements are in place—viewpoint, ostensible repetition of *fabula*, "factual" versus "counterfactual" comparison, metaphysical (or unphysical) world-synthesis—whether or not the story ensues as it is supposed to.

What we may then observe is not the trite conclusion that some narratable stories obey logic and some don't—which will no doubt be a useful guideline for philosophers and historians if time travel to the past is ever invented—but rather the more crucial problem that narrative viewpoint is what both causes and perpetuates paradox in the first place, that it works both for and against the postulate of *fabular* apriority. "Yesterday's Enterprise" may be a poor story by logical standards, but it is nonetheless a fine metaphysical inquiry into the counterfactual viewing position through which all narratives, including paradoxical ones, are constructed. Most particularly, it is a fine speculation on viewpoint itself, which it observes both implicitly, through the presentation of the two shifting realities onscreen (aided by special effects), and explicitly, through the unusual narratological capacities of one particular character, Guinan, who steps in to solve the episode's dilemma.

What Guinan (Whoopi Goldberg) knows about the history of the *Enterprise-D* is also, as it turns out, what the viewer knows: that the course of events currently represented on the television screen is wrong, and that the original timeline needs to be restored, the narrative rift healed. Unlike in "Time's Arrow," it is too late to prevent paradox—the dead Tasha Yar already lives again, alongside other less drastic anachronies—so the need here is neither strictly a logical nor a physical one, but is rather a metaphysical one, or we could even say a generic one, provided the sense of "genre" here is given a particularly strong deterministic valence. In all available senses except that of logic—metaphysical, narratological, typological, and industrial—we, along with Guinan, understand what should and shouldn't have happened, and what therefore must happen *now* in order to set things right: "That ship from the past is not supposed to be here," she tells Picard. "It's got to go back."

Guinan, being still a character and not a viewer, let alone a professional narratologist, expresses the need for conservation in a symptomatic matter, almost in the form of a psychological compulsion, a mirroring, on the level of something like the characterological unconscious, of the generic pressures that tend to compel paradoxical subversion back toward resolution. In the following conversation she trumps Picard's positivistic demand for evidence with what the viewer understands to be a more proper intuition of narrative rightness, her far-seer's insight about the correct postulation of *fabular* apriority:

> Picard: I need more!
>
> Guinan: There is no more. I wish there were. I wish I could prove it, but I can't. . . .
>
> Picard: Guinan, they will die moments after they return. How can I ask them to sacrifice themselves based solely on your intuition?
>
> Guinan: I don't know. But I do know that this is a mistake. Every fiber in my being says this is a mistake. I can't explain it to myself, so I can't explain it to you. I only know that I'm right.
>
> Picard: Who is to say that this history is any less proper than the other?
>
> Guinan: I suppose I am.

In plot terms, Guinan's extraordinary narratological clairvoyance is a *deus ex machina*, an unduly convenient prompt for Picard to do the right thing and send the *Enterprise-C* back through the rift. No character in a realist narrative ought to be able to distinguish a proper from an improper narrative line, any more than Lewis's Tim, lacking Guinan's extraordinary intuition, ought to be able to understand why he "cannot kill Grandfather." Indeed, in Lewis's anecdote it is rather Lewis himself, in his capacity as metaphysical observer, who plays the role of a Guinan, informing us, as though we were the stubborn narrative positivist Picard, that Tim is strictly incapable, that one of the two 1921s is improper, and that the story *should* (therefore) be told in only one particular way. Likewise, Guinan is a proxy Lewis in "Yesterday's Enterprise," reminding us that there should be only one story, not two, and that the postulate of *fabular* apriority must trump the radical freedom of *sjuzhet*. The predictable result of Guinan's intervention is that the present finally does shift back, the proper original *fabula* of the Narendra III conflict and its entailments gets restored, and no one (except Guinan herself, as we are made aware through a brief knowing smile at the episode's conclusion) is the wiser. The alternate history we have just viewed will have been a mere fiction—told in the *sjuzhet* and viewed in the text and paratext, but then un-told, un-represented, the mere dialectical negation of a proper history. Moreover, its elimination represents the collusion of a number of related restorations: a logical one (Tasha Yar will have died in the past), a generic or ideological one (the *Enterprise-D* will have been, in Guinan's words, "not a ship of war [but] a ship of peace"), a television-industrial one (Denise Crosby will again retire from the *Star Trek* franchise), a narratological one (the classical plot with which the episode commenced will be the one with which it concludes), and a "historical" one (the Jamesian *fabula*, sacrosanct and "past" despite the stresses to which this episode has subjected it, will ultimately be conserved and repostulated).

So, although Guinan represents a conservative *deus ex machina* on the level of plot, in narratological terms she signifies something more complex or multiplex. Like Delany's Jewel, Guinan is the viewer's own proxy in the text, employing the viewpoint-over-histories required both to contrast alternatives and to return them to proper arrangements. And like Jewel, she does something that no mere character could do, reading or viewing the textual and paratextual diagramming of the relation of *fabula/sjuzhet* that the television episode, and even the entire series, itself depicts. The most adequate way to describe Guinan's viewpoint is therefore as quite outside the *sjuzhet*, or outside any particular *sjuzhet*. She is the traveler across lines whose activity is structurally required by the episode's paratext, in order to patch the *fabula* back together before its (and the viewer's) hour is up.

We may note, finally, how the propriety of the narrative that Guinan helps to recoup coincides with other, less comfortably formalistic "natural" orders in *Star Trek: The Next Generation*. Undoubtedly, our return to the proper *fabula*, in which the future is a peaceful continuation of something like the present-day cultural mainstream, entails a quasi-orientalist, possibly racist presumption that everything outside of that convenient conflation of ideological norms we call "the Federation" will be another race, another history, another kind of moral choice, to be indulged or speculated through narrative but ultimately to be eliminated via more "natural" *fabular* restoration. Guinan herself occupies an ideally liminal position in this regard. She is the outsider; the celebrity extra in the cast, played by a recognizable cameo actor.[52] And she is the "magical negro" whose wise and exotic intuition inserts itself, from the fringes of the *sjuzhet*, in order to restore the comfortably unselfconsciousness hegemony of what me might call twenty-fourth-century multicultural American life.[53] In these senses, Guinan perpetuates the hard-nosed but maternalistic, and ultimately stabilizing, role often foisted on female black characters (and actors) in Hollywood-style drama, from "Mammy" in *Gone with the Wind* to the television mothers, housekeepers, and nannies of *The Cosby Show*, *Gimme a Break*, *Family Matters*, and *Everybody Hates Chris*. What Guinan knows, in her homely yet esoteric wisdom, is what we readers or viewers of popular culture always already know about mainstream representation. Her surrogate outsiderness in the plot is a reassuring reminder of our own culturally savvy, liberally tinged, vaguely colonialist positioning, both over the plot and over the viewpoint of our proxy within it.

In brief, here the restoration of "natural" *fabula* in the time travel paradox story, its conservation, entails a viewpoint-over-histories that is basically ideological, its position condensing a variety of overdetermined registers: narratological, cultural-historical, political, generic, industrial, even ethnic and racial. Most crucially, the forces of "identification," whatever they may be, continually assist in conserving what must remain second nature for such generic literature or culture. Our viewpoint is *directly* represented in the story by Guinan, whose (in the end) unproblematic intuition about what is right and proper reassures us that no stray logical, psychological, or ideological tendrils will remain dangling. The viewpoint-over-histories that time travel paradox stories open up is shut down fully, its seemingly willful tendency toward alterity, otherness, or subversion foreclosed, at precisely the moment the plot itself concludes, the episode ends, and the proper cast, mission, and essential goodness of the characters are reaffirmed. Any residual knowledge of some other version of the story—a warlike Picard, or for that matter an evil Spock or Kirk, a transporter-duplicated Riker, or a

Romulan Troi[54]—is reduced to mere dialectical opposition, a consciousness of alterity that might have been, but objectively never was; even Guinan's final smile reassures us of its unproblematically straightforward nonbeing. We see similar outcomes in any number of films or television productions in which alternate histories are presented to the viewer as simultaneously counterfactual and actual chains of narrative that end up expunged from the *fabula* after the fact ("now it didn't ever was there," China Miéville writes, aptly wrenching the language).[55] Sometimes, as in the *Twilight Zone* (1980s) episode "Profile in Silver"—a typical revisionist scenario in which Lee Harvey Oswald is prevented from assassinating President Kennedy—such alternatives are offered as a legitimate historical record, only subsequently erased by the retrospective action of time travel; the recent film *Déjà Vu* performs a similar retro-eradication.[56] At other times, as in *Frequency*, *The Butterfly Effect*, or *Source Code*, the alternate that is constructed but then erased by characters' anachronistic intervention (usually wholly in keeping with genre demands) is rendered in the more transparently textual form of a montage sequence, tinged with an oddly proleptic nostalgia for something *newly* relegated to a mode of what "might have been"—perhaps something like what a character in one of Andre Norton's novels calls "photographing the past."[57] In both types of examples, (narrative) conservatism remains the governing trend.

I might recommend a relatively less conservative moment in popular television, the brilliant but much-loathed episode of *The Simpsons* entitled "The Principal and the Pauper." In this episode, it is revealed that the long-time character Principal Skinner is in fact an imposter with the ambiguously ethnic name of Armin Tamzarian, a revelation that throws into question basic ideological tenets that the series both indulges and lampoons. A kind of quasi time travel plot ensues: an alternate history of Skinner is first depicted in full realistic (or *fabular*) detail, then arbitrarily postulated out of existence. Unlike in "Yesterday's Enterprise," no Guinan-like proxy is provided to shield the viewer from the potential subversion of a *fabula* gone awry, and of its threat to the paratextual consistency of the *Simpsons* series overall. By the end of the half hour, either the specific *fabula* of this episode or the series' overall historical continuity is rendered impossible—they are strictly incompatible with each other—and no conveniently intuitive clairvoyant steps in to let us know which postulation of apriority will be left to stand. Instead, the episode ends with another character, a judge, declaring—and this is an astute commentary on the judicial register on which such narratological fiat would necessarily occur, were it possible without severe irony—"I further decree that everything will be just like it was before all this happened, and no one will ever mention it again, under penalty of torture!"

Viewpoint-Over-Histories in "All Good Things . . ."

I want to look at one more time travel story from *Star Trek: The Next Generation*, a story that more subtly or latently explores the viewpoint-over-histories that time travel narratives literalize for us, and the ways in which narratives tend to naturalize, for the benefit of such a viewpoint, their logical and generic closures. In this case it is the chief protagonist himself, Jean-Luc Picard, who comes to collude with our own viewpoint-over-histories, but who achieves that perspective not via a Guinan-like extralinear perception but through the phenomenological experience of story itself: he becomes the "viewer" of the *fabula* without whom it cannot be "read" and restored. I consider this example, therefore, something close to a complete (or fully claustrophobic) illustration of potential paradoxical subversion, narratologically and generically reconstituted and conserved.

In the final double episode of the entire *Star Trek: The Next Generation* series, entitled "All Good Things . . . ," Picard is forced by his sometime nemesis, the omniscient being Q, to flit uneasily between three time periods: his own past (which also corresponds to the series' introductory episode, "Encounter at Farpoint"), his present life onboard ship, and his future retirement in France, suffering from an Alzheimer's-like disease called Irumodic syndrome. The plot revolves around Picard's discovery, in each of these three times, of a "spatial anomaly" that "grows larger in the past" and that seems to require Picard's intervention in order to heal it. As Data explains, in his characteristic role as macrological expositor, this spatial anomaly is a "rupture in the spacetime continuum," the effects of which proliferate backward in "antitime."[58] Thus Picard has directly encountered, and can now view on the ship's own literal screen, the mechanism of a *fabular* paradox. Indeed, it will turn out to have been his own action, in attempting more closely to observe the rupture with the ship's instruments, that is its very cause. Its eventual effects (which in terms of "antitime" occur in the most distant past) are catastrophic. During a brief jaunt to "Earth, France, about, oh, three-and-a-half billion years ago," Q informs Picard that the anomaly "fills your entire quadrant of the galaxy," thereby disrupting the beginnings of terrestrial life. Picard himself, by looking too closely, has paradoxically commenced, in the future, the very anomaly he is trying to repair, and has retroactively "destroyed humanity" (along with, one is left to suppose in passing, all other life on Earth).

We ourselves are abruptly shown Picard's time-shifting in the midst of what initially appears to be a scene of calm exposition. Picard sits drinking tea with Counselor Deanna Troi, discussing the disturbing events he has recently experienced:

Troi: Have you considered the possibility that this was just a dream?

Picard: No, no, it was much more than a dream. The smells, the sounds, the very touch of things. Much more real than a dream.

Troi: How long did you spend in each time period? Did it seem like minutes? hours?

Picard: See, I can't say. At first I had a sense of, a sense of confusion, disorientation. I wasn't sure where I was. And then all of that passed, and it all seemed perfectly natural, as though I belonged in that time. But, I can't . . .

Instantly, the camera cuts from Picard's face, suffused with his effort to recall these disturbing events, to a close-up of an unidentified hand grasping a vine; from the hand, the camera draws back to show the face of a much older Picard, white-bearded, standing in a vineyard, his expression of concentration similar to the one from which we just cut away. Yet quickly the face of the older Picard grows placid as he settles into the task of tying off what are apparently his own grapevines, as though he had been doing it for years rather than (from the point of view of the camera and of the television audience) only a few seconds.

The cut is cleverly made, first focusing on the tactile impression of the hand on the vine, and only then showing the facial expression that permits the viewer to understand that time travel, not mere flashforward or flashback, has occurred. Picard's phenomenological immersion in this future

Star Trek: The Next Generation, "All Good Things . . . ," Picard time traveling

moment is resonant with a lifetime of memories and habits, and with (to himself) a seemingly organic chronology of events leading up to his own present. Thus he very rapidly transfers his sense of immersion out of the prior and into the new milieu in which he finds himself, almost instantly adopting, as it were, the chronicle into which he is thrust. In effect, he is inhabiting what Heidegger names a "world," and his sense of reality is regrounded immediately via the myriad sensory details of his "being-in-the-world."[59]

Of course, I am describing the scene from the perspective of someone who sees the effects of this phenomenological shift from a distance, which is to say, the non–time traveler, or the television viewer. We ourselves are not immersed, but rather observe from above or outside (from our viewpoint-over-histories) the shift that has occurred between times, viewing therefore with a certain multiplex superiority. However, as Picard's temporal jumps increase in frequency over the course of the episode, he too acquires a modicum of this superiority and becomes better able to reacclimate to his "new" times, either reimmersing himself more rapidly or realizing when he has shifted and deliberately refocusing his conscious attention. Eventually, Picard's ability to acclimate becomes nearly instantaneous, and he switches from past to present to future milieux without skipping a phenomenological beat, adapting himself to each circumstance in order to manipulate the minor metaverse that Q has permitted him temporarily to occupy. Indeed, like the fantastically powerful heroes of other metaverse and multiverse fictions, Picard appears able to "move about and select whatever dimensional coordinates he might wish, pick out a spot in the continua, . . . place himself there."[60] In a sense, he is driven to become like the film camera itself, which "must now, in candor, acknowledge not its being present in the world but its being outside its world."[61] And the camera, in its turn, conforms to Picard's virtuosity by matching with increasing formal smoothness the actions and dialogue of his past, present, and future versions; as he shifts, there are fewer and fewer stumbling halts in his speech or movements, fewer reflective pauses before he reacts—hyperspacetime itself becomes his phenomenological "world." Thus Picard finally arrives, through the progress of the plot, at the same viewpoint-over-histories that the viewer had been granted in advance through the filmic apparatus. In essence, he becomes a narratological agent vis-à-vis his own narrative, a proxy for the necessarily synthetic viewpoint of the television audience, not unlike (but more fully cognizant than) Guinan, with a knowledge of proper and improper histories that no character still strictly within a plot ought to possess.

By the end, Picard is something like an amalgam of Tim and Lewis, occupying the structural positions of both the character within the story and the speculative philosophical overseer whose prerogative it is to adjudicate the logical possibility or plausibility of this or that plotline, based on the compossibility of events seen in retrospect. Indeed, as Picard adjusts to his new transtemporal dexterity, he comes to treat himself like a narratological parasite or "rider" in his own body or bodies. Increasingly self-conscious, he no longer shifts fully from one age to another, no longer bothers to re-immerse himself in the various times, but instead remains a singular, subjective meta-Picard, moving between the lines along with the viewer and jumping in and out of the "worlds" of his older and younger selves. And, as successive movements finally converge onto a unified multiplex viewpoint, one that can travel readily across all the temporal differences that the *sjuzhet* provides, differentiation itself looks increasingly more like mundane spacetime, a here and a there that might be located on any map or diagram and visited at will:

> Data: The anomaly is being sustained by the continuing tachyon pulses in the other two time periods. I suggest shutting them down.
>
> Picard: The next time I'm there, it's the first thing I'll do.

Phenomenological immersion, for instance in the world of the most elderly Picard, is now altogether sacrificed to the radically unimmersed perspective of the viewpoint-over-histories, which Picard himself fully inhabits. In the end, from his virtually paratextual position—Kim Stanley Robinson calls such an avatar a "universe body"[62]—Picard reviews the entire plot and solves its riddle, preventing the anomaly from beginning in the future and propogating into the past, thus "sav[ing] humanity."

In effect, Picard the character, initially bound to his simplex *fabular* history, has risen out of this immersion to become what the television viewer has always already been, in collusion with *Star Trek*'s fans, producers, advertisers, and executives—Picard-the-viewer of all the plot's potential divergences from *fabula*, and of all the aspects of its necessary restoration to propriety. He has succeeded, as it were contractually as *Star Trek*'s chief star, in conserving both his own personal history as a protagonist and the *fabula* of that entire (future) history which *Star Trek* constructs and preserves over its several decades. Because we have arrived at the ostensible end of the whole franchise, Picard's act of conservation is permitted to resonate with all the melodramatic flair of *Star Trek*'s reverence for "humanity" and its future prospects. Q addresses Picard one last time in the hyperspacetime they occupy together, where one may avow such extravagant philosophical clichés as the following:

> Q: We wanted to see if you had the ability to expand your mind and your horizons, and for one brief moment, you did.
>
> Picard: When I realized the paradox?
>
> Q: Exactly. For that one fraction of a second, you were open to options you had never considered. *That* is the exploration that awaits you. Not mapping stars and studying nebulae, but charting the unknown possibilities of existence.

In essence, Picard has saved not just "humanity" but, more important, the entire *paratext*, the *show itself*, along with its penchant for exuberant profundity. "The sky's the limit," Picard declares in his final line of the series. Indeed, what *Star Trek: The Next Generation* depicts at its end is a viewpoint-over-histories so encompassing that all galactic history may be seen from there, much in the same way (albeit with less irony and more grandiosity) that, in Rucker's *Master of Space and Time*, Fletch and Harry view the "tangled egg" of their own universe from superspace. Such is the infinite transcendence afforded to narrative viewpoint once it is detached from the immanence of specific characters or positions within a *sjuzhet*, for instance by a paradox story—"whatever it is that's watching, it is not a human," as Philip K. Dick remarks of such supersurveillance.[63] And such is the structural position, here literalized by Picard and Q in superspace, that is capable of viewing an entire universe of potential *fabulas*—even an entire universe of paratextual representations of those *fabulas*—and yet elects to return to the somewhat tawdry "real" one postulated by the narrative itself, and by its mainstream producers and marketers.

SIX

Oedipus Multiplex, or, The Subject as a Time Travel Film:
Back to the Future

My concluding chapter offers a detailed study of a very well-known time travel film, Robert Zemeckis's *Back to the Future* (1985).[1] This film may represent a paragon of time travel storytelling in the mode of straightforward popular literature, indeed possibly the most generic and stridently mainstream version of this peculiar and idiosyncratic subgenre of science fiction. If my basic hypothesis in this book is correct, that time travel fiction is a narratological laboratory, then *Back to the Future*, in its simultaneous straightforwardness and narrative convolution, ought to be an ideal test case.

To begin: the film combines two of the central historical trends of time travel fiction, along with the theoretical problems they generate in their wakes. The first trend is time travel's role in framing utopian social critique, which in *Back to the Future* reemerges in the largely parodic form of a hyperbolic juxtaposition of past and present American life (1980s and 1950s), a sanitized dialectical contrast milked for its slapstick potential. It is the past milieu that is most obviously positioned as utopian here, not the present, although, as I will discuss, many of the film's clichés about historical difference unreflectively reconstruct its present-day 1980s culture as a kind of ideal status quo. Either way, whatever is good or bad about the past or the future is made very obvious. As Robert Zemeckis says, he "wanted to write a time-travel story where you didn't have to know anything about history to enjoy it,"[2] and numerous other critics have pointed out that such reductionism underlies a great deal of American popular culture during the 1980s. In *Back to the Future*, social critique is also filtered through a self-consciously hackneyed notion of the American fascination with "pulpy" science fiction, comically reduced to deliberately overdrawn icons: the eager teenage protagonist, the retrofuturistic automobile, the absurdly convenient nuclear-powered device ("flux capacitor"), the old-fashioned eccentric inventor, and so on. Nonetheless, the residual traces of a prepulp utopianism in *Back to the Future*, the film's oblique or even unwitting allusions to the themes of social progress and political evolution that provoked the earliest uses of time travel in nineteenth-century utopian romance, remain, alongside these bits of previous science fiction and other adolescent literatures, among the variety of generic snippets the audience is invited to consume.

The second trend of the history of time travel fiction the film usefully takes up is the depiction of structuring conditions of narrative in plotlike and diagrammatic forms, the kind of depiction I have been analyzing in the "multiverse/filmic" phase of time travel media. Indeed, this tendency, in which acts of observing or "reading" multiplex narrative lines are immediately embodied by characters and their viewpoints, is carried to a fruitful extreme by a film like *Back to the Future*. The viewpoint-over-histories that time travel stories simultaneously problematizes and literalizes is represented very directly, stripped down to something close to its most consumable possible version, and then virtually merged with the audience's act of sitting in a theater (or in front of some other screen) watching a mainstream Hollywood film. In this sense, *Back to the Future* is less another example for theorizing time travel in popular literature or culture than a decisive accomplishment of that theorization itself, a "popular philosophy of narrative," as my book's subtitle proposes, in which the viewing act—as it were, the paratextual situation of the moviegoer sitting and watching Michael J.

Fox et al. enact the time travel plot—is conceptualized and portrayed in or as the very film.

Ideology as (Multiplex) History

As *Back to the Future* begins, Marty McFly (Michael J. Fox) is living the life of a typical 1980s white suburban teenager, replete with electric guitar, skateboard, and appropriately clean-cut girlfriend, Jennifer Parker (Claudia Wells).[3] The scenario exhibits a kind of ambiguous decadence, exemplified by the contrast between Marty's '80s hipness and the retrograde nerdiness of his immediate family, who will play an important role both in the plot and in the structuring of the film's moral message. There is Marty's geeky, weakling father, George (Crispin Glover), his disheveled and alcoholic mother, Lorraine (Lea Thompson), and a similarly disappointing older brother and sister. Since Marty's parents and siblings are so ill equipped for their familial roles, a more adequate model is provided in the person of Doctor Emmett Brown (Christopher Lloyd), a genius-inventor and scientific iconoclast on the model of Thomas Edison or possibly H. G. Wells's Time Traveller, and an obvious father-surrogate for the ineffectual George. All of these characters are updates of stock figures from older subgenres of science fiction adventure: the smart but impetuous boy hero, the family members whom the hero must outgrow (or grow to understand anew), the older but still young-at-heart scientist-mentor, the idealized and therefore characterless love interest, and, finally, the unkillable but ultimately ineffective nemesis, Biff Tannen (Thomas F. Wilson), who reappears throughout the film and its sequels at different ages and in various incarnations.

Following an extended staging of Marty's family and of his inevitable dissatisfaction with them, Marty joins Doc Brown in the parking lot of the Twin Pines Mall, where Doc is conducting an experiment with a time machine built into the chassis of a DeLorean automobile. Their conversation is interrupted by terrorists—these are "Libyans," sporting kaffiyehs and dangerous scowls, in accordance with the film's rudimentary politics—from whom Doc has stolen the plutonium required to power his time machine. The Libyans shoot Doc, apparently killing him, and then chase Marty, who, while fleeing in the DeLorean, is accidentally transported back thirty years to the Hill Valley of his father's and mother's adolescence. Having arrived in 1955, Marty soon encounters the youthful versions of George, Lorraine, and Biff and commences observing the historical and psychological foundations that will have underlain, thirty years hence, the peculiarly circumscribed '80s suburban world in which he (will have) lived.

Back in the 1950s, Marty very soon learns that George has a desperate crush on Lorraine, upon which, however, he is too shy or inept to act. What takes place next is the film's initial foray into what Constance Penley rightly calls "a primal scene fantasy," during which Marty is "on the scene, so to speak, of [his] own conception."[4] According to a family legend that Marty has heard many times before (and which we ourselves recall from an expository conversation earlier in the film), George had one day been hit by Lorraine's father's car, brought inside the house, and nursed back to health by Lorraine, a sequence of events that initiated their courtship. However, as a result of Marty's anachronistic intervention at this exact moment in 1955—in an echo of *Star Trek*'s "The City on the Edge of Forever," Marty shoves George out of the way and is hit by the car instead—the narrative is disrupted, or rather forked, and Marty now replaces his teenage father as both the victim of the car accident and the consequent object of Lorraine's attentions.

In the previous two chapters, I argued that time travel stories tend to conserve their moments of potential narrative subversion, a conservatism that in mainstream texts very clearly colludes—for instance, in the case of *Star Trek: The Next Generation*—with a broader tendency to clarify or cleanse the text for popular consumption, whether by watering down its potentially inflammatory ideological tendencies, or merely by assimilating them into larger, more cohesive narrative viewpoints. In this vein, we might dwell for a moment on the curiously bland political humor of *Back to the Future*, which essentially treats the latent critical potential of time travel stories, their "cognitive estrangement" of social norms or differences, as just more fodder for the distracted viewpoint-over-histories that the film audience is invited to construct for itself. For example, a short while after his arrival in 1955, as Marty is being tended to by the flirtatious Lorraine—who of course doesn't realize that he is her future son—she calls him "Calvin Klein," because, having brazenly removed Marty's pants while he was unconscious, she mistakes the 1980s brand label on his underwear for a nametag. This moment of misrecognition, like a number of others in *Back to the Future*, merges two of the film's central themes, the Oedipus complex and the suffusion of American life by corporate advertising, into a joke that breezily shrugs off the conceptual and political weight of both. Each theme relies upon a structural premise underlying all of the film's comic juxtapositions of '50s and '80s life: that nostalgia and futurism are interchangeable, or, in short, that the more things change the more they stay the same, an ideological correlate of *fabular* apriority. Thus the film presents American cultural history through a series of minor puns and pratfalls, a catalog of conspicuous but trivial material changes that in themselves imply precisely no spiritual or essential change: the invention of twist-off bottle tops, the distortion of electric guitar music, the ascension of a movie actor to the presidency, the advent of post-1960s slang such as Marty's tagline "Heavy!," and so on.[5]

The jokes aren't always so insouciant as the Calvin Klein gag. While sitting at the counter of the local soda shop, coincidentally alongside his future father, George, Marty briefly encounters Goldie Wilson (Donald Fullilove), an African American busboy who, as Marty "recalls," will eventually become mayor of Hill Valley. Goldie proceeds to scold George about his failure to stand up to Biff Tannen and his gang:

Goldie: Say, what do you let those boys push you around like that for?

George: Well, they're bigger than me.

Goldie: Stand tall, boy, have some respect for yourself! Don't you know if you let people walk over you now, they'll be walking over you for the rest of your life? Look at me, you think I'm gonna spend the rest of my life in this slop house?

Lou Caruthers [the shop owner]: Watch it, Goldie.

Goldie: No sir, I'm gonna make something of myself, I'm going to night school and one day I'm gonna *be* somebody.

Marty: That's right, he's gonna be mayor.

Goldie: Yeah, I'm—Mayor! Now that's a good idea. I could run for mayor.

Lou: A colored mayor, that'll be the day.

Goldie: You wait and see, Mr. Caruthers, I will be mayor. I'll be the most powerful mayor in Hill Valley, and I'm gonna clean up this town!

Lou: Good, you can start by sweeping the floor.

Goldie [to himself as he walks away]: Mayor Goldie Wilson, I like the sound of that.

Goldie is a minor character, and his brief appearance merely provides one of the film's numerous sociocultural juxtapositions between the '80s and the '50s, in this case focused on the implausibility in 1955 (and, presumably, the contrasting plausibility in 1985) of a "colored" mayor of a California town.[6] The film's message at this moment is clear: in the 1950s, African Americans are sweeping floors, but by the 1980s they will be running governments. Or more precisely—and here is the film's characteristic reinterpretation of historical difference as timeless truism—an African American with gumption ("Stand tall, boy, have some respect for yourself"), and given a little timely prodding, will hardly be inconvenienced by anything like the ongoing history of American racial discrimination, nor for that matter aided by the incipient civil rights movement, each of which is quickly elided in the simple dialectical juxtaposition of these two moments thirty years apart. Furthermore, the characters and dialogue in this vignette are so overdrawn—Fullilove literally shuffles through his lines—that the audience (at least the white middle-class audience who is the film's prime demographic, judging from scenes like this one) is clearly meant to *understand* the scene as no more than a silhouette of American racism, inoffensive by virtue of the very blatancy of its reductionism, which is to say, by virtue of the very same attributes that make it offensive.

Aside from the affront of presuming the time-traveling intervention of an unwitting white 1980s teenager (Marty/Fox) as a catalyst for American racial equality,[7] the more basic problem of the film's politics reemerges in these scenes in the very obviousness with which its ideological positions are constructed for the viewpoint-over-histories that receives its social commentary. Few viewers will have missed the comedic point that Marty goes about accidentally transforming stereotypes into social successes, any more than viewers could miss the momentary titillation and subsequent comic annulment of the Oedipus complex that come about through Marty's abortive liaison with his mother. Most important, the film itself does not draw lines of significance between its various caricatures of nostalgia or progress, for instance to distinguish the topic of racial segregation and oppression from, say, the young George's terror at hearing Eddie Van Halen on Marty's Walkman, or Lorraine's misreading of Calvin Klein underwear.

The point is, everyone gets the jokes and gets them thoroughly. Very little of the humor or the ideological simplicity or the often absurd narratological convenience of these various plot devices, not to mention the cute historical relativism invoked to place products such as Pepsi or Toyota in the film, will have escaped even the most casual audience member. Just so, the relative sexual inadequacy of the teenage George alongside the time traveler Marty, the ironic contrast of Lorraine's teenage libido with her future prudishness, and, in general, the hyperbole of the moral comparisons drawn between the middle and the later twentieth century are all fully accessible within the broad humor the film offers up. Again and again, *Back to the Future* invites its audience to draw critical contrasts between the ideological pretension implied by its borrowed stereotypes of American culture and the film's own lighthearted dismissal of their significance, and, in turn, to coordinate the interplay between institutional, economic, and generic forces into a kind of slapstick dance. For instance, Marty's first social encounter in the past takes place in that same 1950s coffee shop (the sequel has an equivalent scene in a future version of the shop nostalgically named "Cafe 80s"; in the third installment it is in an 1885 saloon):

Lou: Are you gonna order something, kid?

Marty: Uh, yeah, gimme a Tab.

Lou: Tab? I can't give you a tab unless you order something.

Marty: Right, gimme a Pepsi Free.

Lou: You want a Pepsi, pal, you're gonna pay for it.

Marty: Well just gimme something without any sugar in it, okay?

Lou: Something without sugar . . . [gives Marty a cup of coffee]

Linguistic misrecognition is a comedic cliché, to be sure, and old fodder for time travel as well as for popular comedy of all sorts.[8] But the particular misrecognition that occurs in this scene contends with a great variety of relativisms: the provincialism of specific cultural eras, the historical arbitrariness of diet fads such as sugar-free soda, the ironic incompatibility of the prototypically postmodern Michael J. Fox with the purer Americanism of a pre-postmodern 1950s, and even the seduction of teenage consumers by corporate catchwords, all alongside a shameless product placement that would seem repudiated by each of these other foci of attention.[9] What we view here is not simple comedic plot, nor simply ideologically inflected or symptomatic representation, but rather the construction, dismantling, and reconstruction of Hollywood genre and marketing strategy itself, from an ironic position that is neither wholly critical nor wholly co-opted, and that needn't be duped in order to be sold a bill of goods.[10] The time travel film at such moments is not simply an example of easy, even lazy entertainment but an (albeit easy) portrait of the *construction* of such entertainment, and an (albeit easy) critique of the relative savoir faire that different historical epochs might cultivate with respect to the blatancy of that construction.

We are watching Hollywood film, in the broadest sense, with all its moral and economic accouterments, being manufactured and consumed; our interlocution here, even our identification, is with the film's production. And the pleasure in such a viewing occurs not within all these relationships per se, nor in the audience's concrete relationship *to* them, but rather in the abstraction out of them, in the "distance of the look."[11] Thus, if anything like identification does occur here—and presumably it does, if pleasure or absorption can also occur—then we identify foremost with the *film*, in its dense cultural and ideological multiplexity, not primarily with its characters or images, all of which are placed in partial or multiple contrasts with one another, and none of which appears simply as object. What I have earlier described as the literalization of viewpoint, accompanied by the physicalization of the *fabula/sjuzhet* relationship in a paratext, is here very direct: we are filmgoers watching a Hollywood movie, one that in turn produces for us both a plot and a certain kind of ideological self-critique. Our viewpoint-over-histories is, rather excessively literally, located in the theater seat or on the couch, in the real spacetime of contemporary viewing praxis.

Primal Scenery and the Production Matrix

We see again the multilayered quality of this "distance of the look," and the paratextual aspects of our viewpoint-over-histories, in the way the film sets out to remove Marty's incestuous dilemma and restore what it perceives, admittedly with a fair degree of wit, to be proper familial relationships: George, not Marty, must take Lorraine to the high school dance and eventually become Marty's proper father; Marty must go back to the future in which he is merely a son, albeit a properly savvy 1980s son. To bring about this repostulation of proper *fabula*, Marty concocts with George a switcheroo of a type canonical in situation comedy since at least Chaucer: Marty will take Lorraine to the dance and pretend to make a sexual advance, at which point George will arrive to punch out Marty and "rescue" Lorraine, thereby committing the (apparent) act of aggression required to reestablish George's dual position: for Lorraine as viable mate, and for Marty as respectable patriarch.

Because all of the film's familial relationships are strictly conventional rather than, in any complex sense, psychological or sociological, the restoration they undergo can occur entirely through turns of plot, and need not confront either actual or fantasmatic erotic desire. Indeed, wherever complicated psychological predicaments would seem inevitable, the protocols of Hollywood PG adventure genre provide simple narrative solutions that allow the film to proceed unimpeded by either anxiety or arousal, protocols that also remain in line with the general conservative tendency of time travel narrative to tie up or forestall dangerous loose ends. Eventually, after a few plot twists, George does aggress against someone—not Marty, as it turns out, but the conveniently intrusive Biff. On the strength of that success, George is then able to take Lorraine to the dance, thereby restoring his family's proper future history, as well as its suitably patriarchal and candidly stereotypical arrangement. The audience keeps track of this ideological repair job through a family photograph from the future that Marty carries with him, in which figures fade in and out depending on how oedipalized the plot is currently in danger of becoming. If George is wimping out, then the future images of Marty and his siblings, whose very existence depends on George's becoming a properly aggressive father, begin to fade from the photo; if George aggresses adequately, the images fade back in. Thus we ourselves monitor the film's progress toward its resolution through the viewing of another viewing act, or, in other words, by watching the film watch itself, and indeed watch itself conserve itself. The film strangely reduplicates a viewpoint-over-histories that corresponds with almost uncanny

Back to the Future, snapshot of the future

precision to our position as moviegoers. The movement in the photograph now becomes simultaneously the metaphorical gauge of temporal alteration and restoration, and the literal embodiment, via special effect, of the text and technology we have paid to see. Both visual acts converge in the "photogrammatic" figure of the uncannily changing snapshot.[12]

When Marty finally does return to 1985, he discovers his family updated and considerably improved as a result of his own intervention thirty years earlier. The revision he accidentally initiated has had the effect of reestablishing his parents' relationship on the basis of male aggression (George's punching of Biff) rather than of mere female sympathy (Lorraine's compassion for the injured George), and apparently this is a more appropriate foundation for a successful 1980s nuclear family. Here the film opts for one of the common modes of conservation within time travel narratives that I noted in previous chapters: the present is restored to what it should have been, and a divergent *fabula* corrected, all with proper deference to the rules of genre. Thus, in this improved 1985, Marty finds his father (as well as his brother) better groomed, more confident, more professionalized; his mother (as well as his sister) slimmer, more vivacious, more supportive. The film's knee-jerk political reaction here colludes with its narratological conservatism, providing a newly pre-oedipal patriarchy at the same time as it invigorates and cleanses its portrayal of the American family through a transfusion of 1950s values: Marty's family now becomes all the more itself, the family it should have been all along, in a quasi-utopian correction back *to* the status quo. Both the ideological and the narratological demands of the film's plot are wholly satisfied by this logically and psychologically absurd ending, which reintroduces Marty into his revised family with barely a hint of uncanniness; as Pauline Kael comments, it "should be a satirical joke but isn't."[13] Indeed, arriving back in this altered future, Marty fits in even better than he did before, since his parents and siblings, previously wimpy, nerdy, or "transgressively slovenly,"[14] have only now, after the time travel revision, caught up with Marty's own hip, mid-'80s savoir faire, in much the same way that '50s music, technology, clothing, and soft drinks will have become properly themselves only thirty years hence, largely in their manifestations as corporate tie-ins.

Let us turn back to what, for this historical revisionism and its conservative resolution, is a primal scene *par excellence*, the 1955 "date" between Marty and his mother, set up by Marty and George for the sake of their switcheroo. Lorraine is sitting in a parked car with Marty outside the high school dance. If the timing doesn't miscarry (as of course it will), George should arrive just in time to "rescue" her from Marty's feigned advance, simultaneously liberating Marty himself from this awkward and potentially risqué scenario. Despite George's predictable tardiness to his own primal scene, however, the film spares itself requiring Marty to sexually assault his own mother by instead having the young Lorraine, who throughout the film remains oblivious to the fact that Marty is really her son, make a presumptive pass at him, unwittingly undertaking the Jocastan kiss at which the film had so far only hinted, but to which it now, slightly surprisingly, commits itself. Having forked the plot again at this juncture, and having headed down the risky oedipal path after all, the film contrives an entirely arbitrary mechanism for unraveling the oedipal dilemma and its threat of incest, but a mechanism still palatable within the PG rating and therefore altogether plausible as a piece of filmic *production*, regardless of its absurdity as *representation*—in short, absurd as *fabula*, maybe even as *sjuzhet*, but still credible as pop-cultural or paratextual artifact. Unilaterally desexualizing the sexual act, while at the same time displacing in advance the significance of the violence that must ensue from that act once George finally arrives on the scene, the mother's kiss simply fails to be either a metonymy or a synecdoche for sexual intercourse:

> Lorraine: This is all wrong. I don't know what it is, but when I kiss you, it's like I'm kissing my brother. I guess that doesn't make any sense, does it?
>
> Marty: Believe me, it makes perfect sense.

It really *doesn't* make sense, either in terms of dramaturgical consistency or of oedipal fantasy, let alone of straightforward teenage lust. Nevertheless, it makes perfect sense to the hybrid generic entity Marty/Fox/Zemeckis/Spielberg, an entity that comprehends perfectly the limits of its chosen genre and here reassures both Lorraine and the viewer, and possibly the viewer's parent or congressperson, of the propriety of Hollywood family-comedic adventure.[15] Marty/Fox thinly veils his generic relief under a comic nod to the confusion of son/brother that the time travel plot has permitted, and to which only he and the viewer, not Lorraine, are privy. His statement "Believe me, it makes perfect sense" is entirely sensible generically or, so to speak, socioculturally: Marty is like a brother to Lorraine, and not just because he is her son. Michael J. Fox is like a brother to Lea Thompson, and not just because he plays the time-traveling Marty McFly. In fact, post–Alex P. Keaton, he may be understood as a generic brother-at-large, an effectively desexualized yet still sufficiently erotic avatar of Reagan-era corporate media and advertising culture itself.

The line "Believe me, it makes perfect sense" is one among several moments in which Marty/Fox/Keaton observes himself as an actor within a kind of movie about himself, or watches himself constructing the plot in which he is acting or is about to act. The multiplex viewer, in turn, watches this complex yet comic act of self-viewing unfold with an explicitness that, in the theorist's terms, resembles cultural critique far more than it does any symptom of ideological obfuscation, let alone of psychological identification. We can suggest, only slightly facetiously, that what is supposed to "make sense" at this moment—that Marty and Lorraine aren't attracted to each other because they are mother and son—can make sense only because they too, like their audience, are watching a Hollywood movie being filmed, and therefore expect the sort of thing that movies like this always do.

The considerable degree of reflexivity within such a moment in *Back to the Future* suggests that it is not representation, primarily, but rather the viewing of representation, or even the viewing of the production of representation, that the theorist must consider in order fully to comprehend the significance of the film's ideological encounter with its mainstream audience. This encounter is something quite different from a symptomatic experience of, or identification with, representations that would be cathected either by cultural or psychological anxieties or, for instance, by any "desire for regression."[16] The *least* plausible motivation we could supply for such a moment is "anxiety":[17] nothing could be less latent or repressed than the motivations for this pseudo-oedipal twist of plot. Yet the encounter remains to be theorized regardless of what may still be seen very easily as the film's extreme yet utterly conventional politics, sufficiently reactionary as to border on crypto-fascism in its glib promotion of the white American boy and the nuclear scientist as epic protagonists, and of the inevitable restoration of a properly chaste mother and family. Precisely what makes the film comic is also what makes it usefully difficult for theoretical analysis: the fact that the viewing act itself, despite its easiness, is not simple or singular, nor even merely multiplex in its assimilation of chronologies, but always *again* reflected, compounded, and reduplicated, continually juxtaposed to itself in slapstick and ironic ways, like a time traveler encountering his or her primal scene over and over again, in comic casualness. Seemingly the entire matrix of the film's production, from plot to actor to genre to corporate backer, is available to watch the film produce itself, and to invite its audience along for the view, precisely to enjoy the critical disintegration of most of the other "symptomatic" viewpoints through which the film could have been experienced as a political or psychological crisis. In this sense, even the viewpoint-over-histories of the filmgoer is strangely incidental or extraneous, since all the critical viewing positions the film makes available seem already to be occupied. The film itself is more subject than object, more viewer than viewed thing.

The Time Travel Film as Subject

What we watch in *Back to the Future* is neither exactly the life and times of Marty McFly, nor exactly the career or shtick of Michael J. Fox, nor exactly the artistic or economic machinations of Robert Zemeckis and Steven Spielberg, but instead a matrix within which all of these agencies are blended, along with soda ads, theme park rides, computer games, and so on, into a complex yet uncomplicated experience. It would be simplistic to call the viewer's acceptance of this interpellation of his or her life by corporate self-production merely some kind of resignation to the commercialization of pop culture. Rather, the shape of experience in the film is such that audience interest and appeal can occur neither in the ideologically obvious form of identification with characters or outcomes, nor merely with some multiplex viewpoint or sociopolitical position in pop-cultural superspace, but rather through an identification (if we can still call it that) with the entire synthetic capitalist-marketing event that is the film in its real produced and distributed being. But to understand this production matrix, which behaves more like a viewing subject than a viewed object, we need a somewhat new model for reading the significance of the common yet complicated reflexivity within a film like *Back to the Future*. The film itself may tell us how to construct such a model, or will at least show us scenes that imply how it is already constructed. At the same time, it may show us the basic structural system that undergirds what I have called the visual or diagrammatic basis of time travel plots, and their accompanying recourse to paratext as the primal medium of narrative conservation. The key will again be to observe how a time travel film is capable, perhaps as no other filmic type, of literally and realistically emplotting the act of viewing.

Late in *Back to the Future*, we arrive at a scene that duplicates a crucial moment from earlier in the film, now shot from a slightly more distant perspective: the mall parking lot where Doc Brown conducts experiments with his DeLorean time machine. As I described before, Marty had departed for 1955 just after witnessing Doc shot by the Libyan terrorists. Now, toward the end of the film, Marty comes back to 1985 intent on rescuing Doc. The plan is for Marty to arrive a few minutes before he had originally left in order to warn his friend about the impending shooting. Here again the effect of temporal revision will be to help restore events and plot to what they should have been in the first place. Although Marty does arrive back in 1985 a few minutes early as planned, he arrives downtown instead of at the mall, and then cannot restart the DeLorean in order to drive to the mall to prevent Doc's death. Running frantically, Marty gets to the edge of the mall parking lot too late, only to witness Doc shot "again," and to witness his own earlier self witnessing the act, before he (the earlier Marty) "again" slips off to 1955 in the DeLorean. In the earlier version of the scene, our viewpoint approximately followed "Marty 1" as he was pursued by the Libyans and escaped in the DeLorean. Now we reobserve the scene from a position one level of subjectivity further removed, approximately following the viewpoint of "Marty 2" as he watches the same events unfold from some distance away. Marty's earlier reaction, seeing Doc shot, was to scream "No!" Now, just at the moment Marty 2 sees Doc shot "again," he is about to react in precisely the same way—this is, so to speak, the reaction Marty *always* has to seeing Doc shot—and his lips begin to form the word "No!" However, at the very instant he is set to utter the word, he hears Marty 1 scream "No!," as though ventriloquized by his own alter ego. Suddenly, Marty is both subject and object in the scene, viewer and viewed. Or more exactly, Marty sees himself enacting the earlier scene; he sees the part of the film that we, the audience, have already viewed. He (Marty? Fox?) sees himself (Fox? Marty?) in the form of a filmic *effect*, the doubling of the movie star that has long been a favorite Hollywood device for showing off filmic technique or technology, and that accords with Hollywood's penchant for heightening the visibility of the celebrity in the starring role. By comparison, we might compare the self-consciously ironic nod with which an elderly Leonard Nimoy, reprising his role as Spock in the 2009 prequel *Star Trek*, finally converses directly with his younger self, at the same time inviting us to observe the *Star Trek* franchise referencing both its own past and the considerable pop-cultural capital its audience has accumulated over several decades of its incarnations. Arnold Schwarzenegger, of course, has made such direct reenvisioning of prior characters virtually into a minor career, from his numerous repetitions of lines ("I'll be back") and props (jackets, sunglasses) in the *Terminator* sequels, to his elaborate burlesque of genre conventions in the perhaps underappreciated *Last Action Hero*.[18]

Where is the camera, our proxy for a viewpoint-over-histories that could reconcile the strands of a multiplex scenography so replete with nods to prior cinema or moments of cinema? Throughout the later mall scene in *Back to the Future*, the camera remains a conveniently invisible observer, at all times one step more distant than the viewpoint of any given character, never an explicit eye.[19] But because, in this particular scene, a moment of

Back to the Future, Marty's "No!"

self-viewing is occurring within the *fabula* itself, any observation the viewer undertakes via the camera is at least duplicated, if not made strangely or comically irrelevant, by the viewing act undertaken by Marty 2, as if all the viewing needed in the scene were already accomplished. Let us now consider what it means for a character or actor within a film—Marty, Michael J.

Fox—to *be* the viewer of that film, not just metaphorically or formalistically, but actually, literally, in that peculiar circumstance only the time machine permits—and for a film therefore to watch itself along with us or alongside us.[20]

In analyzing the basic relation of *fabula/sjuzhet*, beginning in Chapter 4, I discussed the reasons theorists tend to view narratives as repeating the events of which they are constructed: "To narrate is to retrace a line of events that has already occurred, or that is spoken fictively as having already occurred."[21] Instances of more explicit repetition within a plot simply foreground the underlying iterability of narrative in general or, for that matter, of any representation. In turn, such explicit moments construct a secondary allegory about structure out of what must then seem to be an already present primary or postulated history, providing an explicit "doubling of a preexistent or supposedly preexistent line of events."[22] The time travel plot, of course, interjects the additional condition that this doubling occurs in the mode of realism, as a "line of events" on the historical level of *fabula*, not merely *sjuzhet*. Thus the person Marty 2 sees is really himself, neither as a formalistic game nor as a new outlook upon his past or future, nor even as some psychological or allegorical analogue (as Ricoeur says, "the hero *is* who he *was*"),[23] but rather wholly straightforwardly, as an actual other in the *fabula*. Time travel then permits us to scrutinize in literalized form what Christian Metz calls the general conditions of "the subject's knowledge" in a film, and the acts of recognition and self-recognition without which no film, at least not in the shape of a representation or a narrative, would be viewable.[24] *I* am watching something imaginary—*the other* in the film is not real—but in another sense both my watching and the film, as well as my reactions, *are* real, occurring in a hyperspacetime context really comprising my act of sitting in this paratextual theater, viewing this screen. In short, "I am myself the place where this really perceived imaginary" is projected.[25] And who, then, am "I"? Marty McFly? Michael J. Fox?

Let me revisit a strategy I used earlier and ask some extremely basic questions of Marty/Fox, questions that would also be possible to ask of ourselves as the film's audience. I am taking the opportunity provided by the time travel plot to sort out the epistemological relationships built into the seemingly easy act of sitting in the paratextual "theater" and viewing this "institutionalized social activity called the cinema."[26] As usual, the questions are much simpler, much more basic, than any inquiry into what Marty or the film overall represents or ideologizes, let alone into what it is a sign or symptom of. However, the fact that these are epistemological and ontological questions rather than aesthetic or political ones may be seen as an indication of the metaphysical register at which theory must intervene in the viewing praxis of the mainstream text if not to be preempted by the text's own carefree self-critique of its ideological commitments.

First question: *How is it we know that the two Martys—and the two scenes in which they appear—are different?* An initial formalistic answer, which by no means yet accounts for the *outré* presence of a time machine in the plot, would be that we know the two Martys and the two scenes are different simply because they occur at different times or spacetimes. What intervenes between them is precisely the *sjuzhet* of the film itself, its diachronic movement, which effectively juxtaposes the two scenes and the two Martys about an hour apart in the text. In other words, only within the "time of the film," its particular arrangement of *sjuzhet*, are the scenes different, since in the time of the underlying *fabula* they are apparently the same: axiomatically, there would be no *sjuzhet* at all without such an additional discursive duration. Of course, we also know that the two scenes are different within the *sjuzhet* because the movement of the camera differs slightly. The "eye" or "I" of the discourse-production alters itself, and a new, only slightly more generalized viewpoint is introduced into the text, indicated cinematographically by the wider shot in which both of the Martys at the mall, viewer and viewed, are made visible. But this latter, quasi-textual differentiation merely reinforces the former one in the *sjuzhet*, for if the camera did not diverge at least slightly, we would be watching the film itself repeat, in the mode of a formal device, and not the story within the film. Indeed, these two varieties of discursive difference—the one provided by filmic time, a diachronic difference, and the other by filmic viewpoint in the text, something like a synchronic difference—presumably must always be present to distinguish any two scenes in a film, or indeed in any narrative. Identifying the differences between moments of a discourse is what we *do* when we watch or hear or read a story, and what it means to be the audience of a narrative.

Back to the Future, new viewpoint at the mall

Hence my second question, the more difficult one: *How do we know that the two Martys, and the two scenes, are the same?* Here is a thornier epistemological dilemma, one that the time travel plot makes especially pressing, since the scene is being replayed not merely as a formal or directorial device but "really" being replayed—it is really happening again in the *fabula*. In light of this realism, formalistic or merely *sjuzhet*like answers to the second question become irrelevant. We want to know not how the two scenes (or the two Martys) are made to *appear* the same by the film, but rather how they *are* the same, just as I might ask, how am I the same person who appears in a photograph from many years ago? Thus the answer to the second question must be primarily psychological, not formal, although not therefore a question of identification or cathexis: we know this is the same scene before us, and the same Marty, because we recognize the same subject and subject's "historical" life within the scene, both despite and because of the new viewpoint or hyperspacetime orientation. And we recognize that this same subject or history would also recognize us, so to speak—would confirm our position of seeing-again and our postulation of the scene's *fabular* apriority. But this must be true of Marty, as well: at the moment he sees Doc Brown shot again and is about to scream "No!," he recognizes this other Marty as himself and observes his own subjective reaction from afar. Simultaneously he reflexively solves the temporal conundrum at the heart of the narrative—what *was* in the scene now *is* again—along with the psychological conundrum—the *other* he now sees is really *himself*, about to become what he has already been. His past is that other past only as a kind of future to which he has now come back, within which he will eventually have re-viewed the history of himself watching Doc get shot. *Fabula*, revisited as real, becomes paratext, becomes *film*, and Marty watches it happen before us.

Despite the drastic complexity of this unfolding temporality, the audience observes easily and immediately that Marty's history compels him to recognize himself sufficiently viscerally to exhort that simultaneously voiced "No!" The story of Marty, at this moment, comes around again with the inexorability of instinct or reflex, even fate, and entirely consistent with the more generic fatalism of classical Hollywood narration. Marty himself, like the film itself, now "is who he was," at almost the identical moment that genre expectations are conserved by the anticipation of the happy ending. Doc will be alive after all. But I have yet to answer my own question: *how* do we identify this identity and this inevitability, the fact that the two Martys are the same? How does Marty do it? What is it, precisely, that gets itself recognized?

Lacan gives us a grammar to consider the question more adequately. In describing the "realization of the [subject's] history" within the discourse of analysis, Lacan reminds us, closely following Freud, that a history, if it is to be significant for the analysand, must be an active and motivated reconstruction, an event in that subject's present development: "[T]he restitution of the subject's wholeness appears in the guise of a restoration of the past. But the stress is always placed more on the side of reconstruction than on that of reliving. . . . What is essential is the reconstruction."[27] In the psychoanalytic session, I travel back and visit my *fabula*, so to speak, but only insofar as I now renarrate and resymbolize that traveling, both for myself and for the other to whom I narrate it, reconstructing it in full speech "in view of what will be." The tenses required to describe this multiplied temporality are necessarily convoluted: "What is realized in my history is not the past definite of what was, since it is no more, or even the present perfect of what has been in what I am, but the future anterior of what I shall have been for what I am in the process of becoming."[28] I am my own history, yes: but neither as something over and done with (the past definite of what I was), nor merely as the chain of events which led to the current "I" (the present perfect of what has been in what I am), but rather as what I now construe and declare will become my meaningful past (the future anterior of what I shall have been) as I continue to become all the more myself into the future (for what I am in the process of becoming).

The way Marty recognizes himself—and the way I, the viewer, recognize the two Martys as the same from my multiplex viewpoint-over-histories—and finally (although this takes the story out of the film, so to speak, and into the theater) the way I recognize Marty as Michael J. Fox and recognize myself as the cinematic audience—is the same way I always recognize the continuity of any subject over time: I renarrate a plot that ties its *fabular* moments together and reshapes these moments into a synthetic identity for the sake of a future narrative, the contours and coherence of which, whether real or fantasmatic, I already anticipate. In essence, the subject *is* that (future anterior) narrative, which both distinguishes and interrelates its multiplex selves through the mechanism of a paratext, respatializing and resynchronizing diachronic differences in the form of particular points of view: *there* is what I shall have been, *here* is what I am in the process of becoming, *now* is when I see the whole coming together before me. What else would it be to recognize oneself over time, but to *film* oneself from two different positions, juxtaposing within a symbolic (or, if one is a time traveler, a literal) space the difference-in-time that made one a synthetic, multitemporalized subject rather than a simple, singular point of view? Marty recognizes himself for precisely the same reason that I recognize the two Martys as the same, and for the same reason we recognize any subject as the same: because he knows how to construct the narrative that connects them and can retell that narrative, or refilm it, at the very moment he is about to scream "No!" and hears his other self do so. He recognizes himself because he, too, has seen the movie, or is now watching it along with us—the movie, or its structurally equivalent multiplex act of viewing, being the necessary paratextual medium that simultaneously differentiates and connects the two scenes and the two Martys, a produced artifact in collusion with the viewpoint-over-histories I come to occupy.

Indeed, time travel films are so fully capable of toying with this relega-tion of *fabula* to textual tautology—the event occurred that way, because I saw it in the film—that a character can, still in the fully embedded mode of plotlike realism, set about conserving "past" scenes of the *fabula* solely in conformity with what has appeared on the level of the filmic *text*. The main character of *Timecrimes* does this, enacting a series of seemingly arbitrary poses and movements simply because he remembers having seen himself enact them earlier in the film (but later in the *fabula*). In *Déjà Vu*, the pro-tagonist, having arrived in the past, arranges some magnetic letters on a woman's refrigerator to read "u can save her" because he recalls having seen them arranged that way in the (*fabula's*) future. The protagonists of *Primer* repeatedly travel back to earlier moments of the story, attempting to reconstruct, with the obsessive fatalism of dramaturges, the precise chain of events that both we and they know to have occurred "already" in the film's plot. Considering such apparently paratextual revisionism, we might consider time travel in film a multiplex surveillance of *fabular* lines, and therefore continuous with the thematic linking of film-viewing and sur-veillance technology that has preoccupied mainstream cinema from Hitch-cock's *Rear Window* through a recent spate of hyperbolic world-viewing fantasies such as *Déjà Vu, Enemy of the State, Source Code,* and *The Adjustment Bureau*.[29] The time traveler, even in so humble an instantiation as Marty McFly, would be an emplotment of this collusion between technology, view-ing praxis, and generic plot construction: a character in whom acts of hyper-spacetime renarration and *outré* technologies of espionage add up, finally, to a matter-of-fact figuration of cinematic viewpoint.

Here, then, in the peculiar crisis of the time traveler, we have the film's true primal scene, in which we discover that *Back to the Future* is not about a particular subject, Marty McFly, but rather *is* a subject, one that watches itself via filmic or quasi-filmic technology, and something both more com-plex and more quotidian than either the character Marty McFly or the viewpoint-over-histories entailed by his paradoxical doubleness alone could be. In fact, to return briefly to Lacan's terms, the film itself is the *analysand*, a storyteller that observes itself traveling back, recognizes itself as differ-ent, and nonetheless comes back to its own future to complete itself inexora-bly in sameness and identity, willing symbolically to return to its own primal scene to watch and narrate it(self) again in future anterior—in a word, to conserve itself. The film is an analysand, and we, the viewer (not the theorist per se), are the recipient of its full speech, its analyst—or would be, if our participation in the analytical reconstitution of Marty's history weren't already accomplished by the film. Therefore the film has all the deceptive straightforwardness of an individual psyche, which could not be more certain of its own completed identity, regardless of how impossibly complex the *mechanism* of that self-identification might be, were it to be adequately theorized.

The Multiplex Camera

We are now in a position to understand fully the peculiar viewpoint-over-histories that time travel fiction literalizes, and that can tell us something about how all narratives are read or viewed. We can continue by noting that the structural positions through which the viewer encounters the film are identical to those through which the film encounters itself. This is a generalized or quasi-ontological condition of perception within the cinema, one that film theorists have tended, far too quickly, to call the viewer's "identification" with the camera, even where they explicitly disapprove of this term.

When I first discussed the scene in the mall parking lot, I tentatively described the camera as something like a Pudovkinesque "invisible observer." But the equivalence I propose between the filming of the time travel repetition and the viewing of it by the posterior subject (Marty 2), and in turn by the viewer, ought to indicate why the ostensible invisibility of the camera's observation must be more carefully scrutinized. Seen from the perspective of Marty 1, the "later" observation of Marty 2 would in fact be like an invisible camera, but that means precisely that this invisibility is not equivalent to just any objective viewpoint, but rather only to a viewpoint that narrates (or narrativizes) the subject's own temporality. In other words, it is the camera, not Marty 2 per se, that is the subject's "eye" in recognizing the two Martys as both different and the same, and therefore as the self-identical Marty. But if this is the case, then it is neither Marty 1 nor Marty 2, nor any combination of them alone, that sees and then *is* the synthetic thing called "Marty," without the additional subjective component of vision supplied by the camera. Subjectivity itself, here, is ineluctably cinematographic, something filmed, a "kind of meta–time travel story . . . *enacted* by the film-viewing experience."[30] To put it another way, the camera—that mechanical contrivance which connects our presence in the physical spacetime of the theater (or on the living room couch, or before the computer or smartphone screen) to the perspective of the film's narration—is the fundamental (paratextual) medium of our requisite viewpoint-over-histories, the real hyperspacetime "chronograph"[31] by which the hierarchical relation of *sjuzhet* and *fabula* is postulated and conserved, and by virtue of which we "fly across duration itself."[32]

In terms of how it constructs filmic plot, the camera that sees identity and difference is, as Deleuze shows us in detail, essentially mobile: "[T]he mobile camera is like a *general equivalent* of all the means of locomotion that it shows or that it makes use of—aeroplane, car, boat, bicycle, foot, metro."[33] The cinematic apparatus is uniquely capable of such a "pure movement extracted from bodies of moving things."[34] Thus the camera travels, so to speak, across the differences of what it shoots—or what is the same, vis-à-vis narrative, across the times—and, in its absolutely essential movement, becomes the thing that brings those differences within a whole, gives them their respective identities. As characterless as Robert Zemeckis's filmic apparatus may appear to be, or as invisible as it may remain to the character Marty at the moment he recognizes himself across the mall parking lot, this movement of the camera behind Marty is the indispensable structural element required of the subjective comparison he is in the process of making, a moment enabled only by the traveling of the "shot" through time: "[T]he shot is not content to express the duration of a whole which changes, but constantly puts bodies, parts, aspects, dimensions, distances and the retrospective positions of the bodies which make up a set in the image into variation. The one comes about through the other."[35] The viewpoint that reconnects Marty as self and other in the instance of time travel is thus neither Marty 1 nor Marty 2, but is rather a narrative moment neither identifiable as, nor strictly distinguishable from, either of the Martys: a partially generalized moment of watching that constructs the "fiction or imaginary principle" of identity[36]—a *general camera*, which films *a priori* a montage of spatial and temporal differences.

Montage, as Deleuze notes, "does not come afterwards."[37] It is the primary form of filmic comparison, precisely because only the actual or potential movement of the camera within the whole of the shot can constitute the kind of viewing the film does, and that we then do, of a differentiated equivalence like that of the encounter between Marty 2 and Marty 1 at the mall parking lot. Thus montage, as Bazin already realized, is a "metaphysical" construction and a meta-analysis of reality itself, a technics that "presupposes of its very nature the unity of meaning of the dramatic event."[38] Deleuze glosses Dziga Vertov's concept of a "kino-eye," then follows with a series of broad claims for the primacy of filmic vision in the construction of the world of subjects and objects in general: "What montage does, according to Vertov, is to carry perception into things, to put perception into matter, so that any point whatsoever in space itself perceives all the points on which it acts, or which act on it however far these actions and reactions extend."[39] The film camera, in other words, represents the general apparatus of perception itself, and postulates its in-principle unmitigated capability. Deleuze reminds us, again quoting Vertov, that "this is the definition of objectivity, 'to see without boundaries or distances.'"[40] And referring to Bergson, who is his more basic source for a model of general ontological perceivability, Deleuze writes that "the eye is in things, in luminous images in themselves."[41] This eye, the eye of matter—which means, of the world of images, which is the *whole* of the world, unallied with any specific seeing subject or seen object (although it may have narrower interests in some of them)—this is the eye that sees Marty 1 and Marty 2 as two different versions of the same self. Here, finally, is our essential psychological generalization about film viewing, which the specific instance of the time travel story makes especially clear: not just that the film is a subject, told or retold from a multiplex viewpoint-over-histories, but rather that the *subject* is a story, a *fabula* about itself over time, renarrated from a *filmic* point of view in hyperspacetime. Therefore, the subject is a story told outside of, or after, the temporal frame in which it itself *exists*, a post-ontological narrative retold in or as a "movie." In essence, *the subject is a time travel film*, a story that travels out of its own time and sees itself "again," only to re-present itself as the (para)text of that re-viewing. Marty doesn't recognize himself *in* the film but *as* the film. We don't recognize ourselves "in" Michael J. Fox, but "as" Michael J. Fox—who is himself, however, already also a viewer watching the movie being produced and screened before us.

In short, the complication we discover is not in the time travel film but in ourselves, just as the straightforwardness of the mainstream time travel film is not in the film per se, but is "in" the film's collusion with our own easy ability to *be just like the film*, not by identifying with it but precisely by being the paratextual mechanism of self-reflection that it both demands and supplies. But of course, as the latter portions of my reading of *Back to the Future* should remind us, an artifact of popular culture *is* still ourselves, and theory gives up too quickly if it considers this merely a cliché or an allegory about the putatively double life of the theorist as both viewer and critic. Ontologically speaking, the film and the culture are ourselves because, first, they literally *are selves*—subjects of a self-viewing discourse that narrates itself to another, a real spacetime audience, in a kind of full speech, and solicits that audience's capitulation to its peculiar formation or conservation of identity and significance. There is, in a sense, no theoretical analysis of the time travel film to be done—only the physical observation of a self-analysis already in progress, to be redescribed by the cultural theorist along the lines of an empirical case study.

Conclusion: The Last Time Travel Story

> It is clear then that there is neither place, nor void, nor time, outside the heaven. Hence whatever is there, is of such a nature as not to occupy any place, nor does time age it; nor is there any change in any of the things which lie beyond the outermost motion.
>
> —ARISTOTLE, *On the Heavens, Bk. I, Ch. 9*

> To see the universe as a single object is a great thing.
>
> —RUDOLF V. BITTER RUCKER, *Geometry, Relativity and the Fourth Dimension*

In my first "Historical Interval," I proposed, of course polemically, that Harold Steele Mackaye's *The Panchronicon* ought to be considered the first time travel story. Both historical and formal reasons compel this suggestion, even if the retrospective judgment of genre history might seem to render it frivolous. Mackaye's novel appears in 1904, at a moment when time travel fiction teeters on the verge of solidifying the narrative and speculative interests that inform its identity as an independent story type, having found itself abandoned by the prior type that had so far accommodated and justified it, the utopian romance. The result is a thoroughly dialogical recasting of the utopian time travel macrologue, and hence a species of metafictional or "double-voiced" text that persistently writes into the plot its own mechanics of self-construction—in a word, parody.[1] That time travel starts as parody—even if there will be numerous moments later in the genre's history when it forgets this origin and takes itself entirely seriously—should not be a surprise. Time travel fiction is a direct confrontation with the processes by which narratives are constructed and presented—it is, in the technical sense used by narrative theorists following Genette, the most *discursive* of popular genres. Thus, if the genre of time travel fiction comes into its own at the very moment it becomes a metafiction about itself—possibly even an elaborate self-deprecating joke—this is all the more its strength as a genre, establishing the reservoir of fruitful convolution and hyperbolic plot manipulation from which it will continue to draw.

I would like now to offer, still with the aim of illustrating essential or ongoing trends in the history of time travel fiction, a few nominations for the last time travel story. This would obviously not be "last" in any strict chronological sense, since time travel fiction and film are generated today as prolifically as ever, but rather in a structural or theoretical sense: "last" as in the most severely self-encircling, the most imaginatively self-duplicating, the most hyperbolically metaversal, or, generally speaking, the most incisively narratological. To put it one more way, the last time travel story might be the most elaborately parodic (even if quite serious) popular text, an assemblage of the modes of extremity to which narrative convolution can be pushed and yet still remain comprehensible as plot. What follows, then, is inevitably an idiosyncratic list, but I hope it may at least point to some provocative moments of narrative experimentation across a range of literary and artistic culture.

Psychology of Completion

At several points in the book, I have described the genre of time travel as a strategic suppression of social and psychological issues, possibly even a screen for such topics, which persist within time travel stories largely at the fringe or in the background. What supplants these issues—in accord with the rise of relativity physics after the 1920s, and then with the consolidation of the technical or technological bias of "hard" science fiction from the 1930s on—is an abiding concern with the *mechanisms* of time travel, and with the innovative storytelling forms that such mechanisms can generate. In this respect, a book like James Hogan's paradox thriller *Thrice Upon a Time*, which is obsessed with describing the mechanics of sending and receiving transtemporal messages—a kind of ultratechnical revision of

Gregory Benford's more self-consciously sociological *Timescape*—could be considered a last time travel story. Another choice might be Stephen Baxter's *Manifold: Time*, Greg Bear's *City at the End of Time*, or the film *Primer*, the last of which possesses the additional merit of depicting perhaps the most strictly realistic milieu for the invention of a time machine yet done—not in terms of physics or engineering, but in terms of setting, character, and dialogue. Finally, looking in a very different direction, one can discover a perfectly "hard" time loop story, complete with a predestination paradox and a consistently rendered closed timelike curve, in an episode of the children's show *My Little Pony: Friendship Is Magic* entitled "It's About Time" (2012)—and with that discovery, reasonably surmise that the popular appreciation of time travel technics has arrived at a certain denouement.

However, a candidate for the culmination of specifically narratological recomplication might already be found as early as Robert Heinlein's "All You Zombies—" (1959),[2] a story that pushes to a certain limit the loop mode of time travel fiction that Heinlein himself helps establish in 1941 with "By His Bootstraps." "All You Zombies—" is a *noir* set piece with four players: a bartender/narrator in a New York City dive called "Pop's Place," a disaffected freelance writer who wanders into the bar, a teenage girl named Jane, and an orphaned baby. As the story twists through a sequence of spatiotemporal shifts generated by the narrator's "U. S. F. F. Co-ordinates Transformer Field Kit, series 1992, Mod. II," we gradually discover that all four characters are the same person. To make the plot cohere, Heinlein must write a series of interconnected subplots, each in a different time, and must also, of course, arrange for his main character to change genders and then both impregnate and give birth to him/herself. This latter subplot proceeds by the fortuitous discovery, on the operating table during Jane's Caesarian section, that she is a hermaphrodite and requires (by the logic of mid-twentieth-century sexology) a sex-change operation.[3] The narrator then sets out to arrange, via the time machine, the series of encounters that will compel Jane and her male avatar to meet and fall in love, and their baby to be kidnapped and installed in an orphanage in the even more distant past. In the end we are shown explicitly what we already anticipate, that the bartender/narrator himself *is* Jane (and also her lover, and also the baby): "A Caesarian leaves a big scar but I'm so hairy now that I don't notice it unless I look for it."[4] The story, despite its moments of characteristically hyperbolic sexism and glibness, ends on this peculiarly moving portrayal of the remoteness of its own self-contained fictional world:

> I *know* where *I* came from—but *where did all you zombies come from?*
> I felt a headache coming on, but a headache powder is one thing I do not take. I did it once—and you all went away.
> So I crawled into bed and whistled out the light.
> *You* aren't really there at all. There isn't anybody but me—Jane—here alone in the dark.
> I miss you dreadfully![5]

Here the narratological micromanagement of "By His Bootstraps" is taken fully to a logical conclusion, regardless of its sociological (and biological) absurdity: a self-perpetuating causal paradox in which no perspective remains outside of the narrator's own immanent worldline, his/her *sjuzhet*. Nonetheless, some such external perspective is hinted at by the story's own oblique reference to the "zombies" who may be said still to inhabit the *fabula*, and who seem to include the reader, now addressed in the second person: "*where did all you zombies come from?*" Are "zombies" those of us who, unlike the time traveler, emerge uncannily out of our own lost or immemorial pasts, essentially defined by the irrecoverability of our own histories? By comparison with "zombies"—which is to say, by comparison with actual, nonfictional human subjects—the uncanniness of the time traveler's isolation consists in the very completeness of his/her own personhood, the absence of any constitutive gap between present and past. Heinlein is no great reader of Freud—indeed, frankly he is not much of a psychologist at all—yet there is a certain profound insight offered here, as it were by virtue of the relation of the *sjuzhet* and *fabula* itself: the "zombie" is a subject both tied to and cut off from its own origins, fallen into belated self-alienation. The removal of such alienation—readily supplied in the time travel story by the narrative shortcut of self-creation or self-authentication—in psychoanalytic therapy would represent a fantasmatic asymptote: its accomplishment would mean the annihilation of adult consciousness or the reversion to the timelessness of the infantile unconscious, in which the past is fully here and now but any perspective from which to *retell* it is missing.

As I suggest in Chapter 6, subjectivity itself may be understood as a kind of "time travel film," an ongoing "story" that retells or re-views such a temporal-subjective difference, continually stitching it back together for the sake of a future narrative of self that necessarily remains unfinished. In that sense, the protagonist of "All You Zombies—," having always already completed his/her own narrative, suffers from the lack of *lack*, an ironically dialectical reversal of the constitutive temporal incompleteness of human subjectivity. The time traveler is a fantastic or fantasmatic being, not primarily because time travel itself is fantastic or fantasmatic, but rather because such a being represents something like the subject's own fulfilled death wish. Being in a position literally to rediscover one's own past/ unconscious—complete with its own inherent bisexuality, its fully unrepressed infantile stages, and its fully visceral awareness of the artificiality of "normal" adult consciousness, what Freud calls the "secondary process"— would be tantamount to the annihilation of every constraint and dissimulation that constitutes the adult subject as such. No wonder we normals, stuck in our immutable and amnesiac linear timelines, would seem like "zombies" to such a fulfilled psychic entity. In psychological terms, "All You Zombies—" presents a realist depiction of a fantasmatic ideal, in comparison with which normal subjectivity is exposed as weird undeath. Its protagonist achieves a temporal or historical closure toward which human subjects strive but which, by definition, they cannot achieve in a life, only in a (paradoxical) plot.

In literary-theoretical terms, not coincidentally, "All You Zombies—" also offers the potential to overturn one of narrative theory's most common shibboleths: the definition of the fictional as "incomplete" and of the real world as "complete." For Heinlein's story, quite rightly, the reverse is true: it is the fictional narrator who is complete, and the real subject (the zombie) who is incomplete;[6] the former's suffering is itself an uncanny dialectical echo of the latter's, or a gloss on the constitutive narrative *gap* of subjectivity, the impossibility, in real life, of simultaneously telling and being, or of existing in/as one's own primal past.

Politics of Reversal

In proposing Heinlein's "All You Zombies—" as a last time travel story, I so far remain solidly within the confines of generic mid-twentieth-century science fiction and its customary post-Einsteinian story type. However, as I suggested already in the Introduction, time travel as a mode is far broader than that; as what I called a "mechanism for revising the arrangements of stories and histories," it potentially represents a fundamental feature of narratives in general, to be observed wherever a writer provides either explicit or tacit devices to manipulate the order, duration, or significance of events.

I have already cited some instances of non-science-fictional narrative experiments that might be called time travel either in their specific confrontation with the significance of temporal ordering or in their experimental construction of narrative viewpoint. All three novels of Beckett's trilogy are apropos here, in particular *The Unnameable*, which takes place in an uncanny hyperspacetime, a physical environment almost like the interior of a protofilmic magic lantern, within which Beckett's own fictional inventions are reviewed by its quasi-transcendental protagonist. Metafictions that thematize the physical process of their own production or narration, the "scene of writing," probably in general occupy the borderlands of time travel fiction, as does experimental writing or filmmaking that reverses time: Kurt Vonnegut's *Slaughterhouse-Five* (1969), with its description of reversed film sequences of bombings; Chris Nolan's film *Memento* (2000), with its reversed shots of the opening murder; and Martin Amis's *Time's Arrow* (1991), which narrates the entire life of a German concentration camp doctor backward from death to birth.[7] In all these examples, temporal reversal is also tantamount to political or moral upheaval; each work contrasts a simplex interpretation of events in classical time with a perspective unsure or ambivalent about the relative significance of these events when "played" backward or forward. Each is thus also a reflection upon the peculiar value a reader or writer invests in the present moment of narrative focus, and the selective devaluation or derealization of precedents and consequences—as it were, a thermodynamics of narrative predisposition.

However, as a last nongeneric story concerned with the politics of temporal reversal or retrospect, I would nominate Toni Morrison's novel *Beloved*, a book that attempts, both in its plot and in its narrative structure, a "working through," in the sense Freud uses this term, of the psychological and sociohistorical trauma of American slavery. *Beloved* commences in a house haunted by a baby's ghost: years before, the main character, Sethe, had killed her child rather than allow her to be returned to slavery. The physical atmosphere of the present-day house is "palsied by the baby's fury at having its throat cut,"[8] and Morrison's novel is most often described as a "ghost story" because of the physical manifestations—which is to say, the not-strictly-psychological or not-strictly-allegorical manifestations—of the haunting of 124 Bluestone Road: "The sideboard took a step forward"; "Paul D . . . followed her through the door straight into a pool of red and undulating light that locked him where he stood"; "the baby's spirit picked up [the dog] Here Boy and slammed him into the wall."[9] Of course, it is equally the narrative itself and the reader's experience that the book intends to "haunt," and so its present distress unfolds in fits and starts, carefully loyal to the traumatically fragmented—but by no means, for that reason, less powerful—effects of the past upon its characters. In essence, the novel's *fabula* cannot be reconciled, can *never* be reconciled, with a coherent, classical *sjuzhet* in its own present day, but rather remains unassimilable in its continual, traumatic influence—a history not forgotten but rather repressed, and therefore, of course, all the more present in its fraught non-appearance for the subjects who (un)remember it.

"In trying to make the slave experience intimate," Morrison writes in a foreword, "I hoped the sense of things being both under control and out of control would be persuasive throughout; that the order and quietude of everyday life would be violently disrupted by the chaos of the needy dead."[10] These disruptions take several forms in addition to ghostly commotion, for instance the abrupt *medias res* of the narration itself, which thrusts the reader into a domestic environment both "under control and out of control" without expositing who or what motivates its disturbing anthropomorphism:

> 124 was spiteful. Full of a baby's venom. The women in the house knew it and so did the children. For years each put up with the spite in his own way, but by 1873 Sethe and her daughter Denver were its only victims. The grandmother, Baby Suggs, was dead, and the sons, Howard and Buglar, had run away by the time they were thirteen years old—as soon as merely looking in a mirror shattered it (that was the signal for Buglar); as soon as two tiny hand prints appeared in the cake (that was it for Howard).[11]

Soon, a medley of occluded or ambiguous images from the past—a chokecherry tree, milk, red light, water, blood, "men without skin"—crowd upon the narrative, in one sense like symptoms of an unspeakable ordeal, but in another sense like remnants of histories already told far too many times, obsessively and probably never adequately, before the present narration has begun. Indeed, somewhere among the novel's backstories is the whole history of the Middle Passage, referenced in Morrison's epigraph—"Sixty Million and More"—a trauma obviously too early and too pervasive for any specific character in the book to remember it directly. The exposition of both supernatural events and repeated images of the shared past unfolds gradually, like a halting psychotherapy, without its ever becoming precisely clear to what degree its etiology is psychological. Characters retell history with their bodies at least as much as with their words, in physical actions and reactions: Denver "stepped into the told story that lay before her eyes on the path she followed away from the window"; Paul D realizes "that his legs were not shaking because of worry, but because the floorboards were."[12] Here is Sethe's own spontaneous theorization, describing time itself as an incapacity to distinguish either between the physical present and the "picture[d]" past or between an individual "rememory" and a social-historical inheritance:

> "I was talking about time. It's so hard for me to believe in it. Some things go. Pass on. Some things just stay. I used to think it was my rememory. You know. Some things you forget. Other things you never do. But it's not. Places, places are still there. If a house burns down, it's gone, but the place—the picture of it—stays, and not just in my rememory, but out there, in the world. . . .
> . . . "Someday you be walking down the road and you hear something or see something going on. So clear. And you think it's you thinking it up. A thought picture. But no. It's when you bump into a rememory that belongs to somebody else."[13]

One motive for such ambiguity between the physical and the psychological is surely the refusal on Morrison's part to distinguish strictly between personal narrative and history. Such a refusal is proper and even unavoidable for either the novelist or the historian of atrocity, since the partitioning of historical fact from psychic effect in such contexts would be precisely tantamount to forgetting, an evisceration of the embodied assemblage of types of "rememory" that grounds any individual life in the present. Thus, although it is the character of Beloved herself who is closest to a literal time traveler in the novel, it is rather the other characters' reactions to Beloved, and to the shared *fabula* they can neither fully describe nor purge, that constitute the novel's most compulsive and effective "traveling."

The book eventually arrives at a remarkable section, about two-thirds of the way through, in which a series of shifts in narrative voice are presented in brief, fragmentary chapters, each seeming to depict a different character's version of the shared *fabula*.[14] Here is a part narrated by the dead child herself, playing on the ambiguity of the name/noun/adjective "Beloved":

> I am Beloved and she is mine. I see her take flowers away from leaves   she
> puts them in a round basket   the leaves are not for her   she fills the basket
> she opens the grass   I would help her but the clouds are in the way   how can I
> say things that are pictures   I am not separate from her there is no place
> where I stop   her face is my own and I want to be there in the place where her
> face is and to be looking at it too   a hot thing[15]

At such a moment, Beloved's quasi-symptomatic, quasi-allegorical return from the past has, so to speak, infiltrated the voice of the narration itself, in a form of speech now detached from either specific character or plot, and thoroughly "achronic," in Genette's term. We cannot decide for certain whether we are reading *sjuzhet*, *fabula*, or text. Most important, our inability to decide is caused by the weight of the *fabula* itself, which returns to the narrative present with all the distorting influence of a primal trauma, one that will simply not permit a devolution into coherent, classical *sjuzhet*. In short, the narrative falls into fragments because what travels back from the cathected past is all too real, even all too *physical*, a burden, inhibiting the construction of plot and instead compelling an unsublated, too-concrete paratext—which nonetheless provides a way to "say things that are pictures." *Beloved* is a time travel novel, and a brilliant one, because it illustrates the relative strength of *fabular* apriority, its constraint—not logical, finally, but fundamentally historical—on what can or cannot be uttered in the *sjuzhet*, and what may instead have to be enacted in physical "rememory" or revisited directly on the page.

Diagrams of Histories

The results of *Beloved*'s confrontation with trauma—its concrete presentation of the language of potent and ambivalent recovery or "rememory"—

bring us again to an idea I proposed in Chapter 4, that one thing time travel stories do especially admirably, even in purely textual forms, is to "diagram" themselves, depicting an essentially visual layout of multiple narrative lines. Samuel Delany's novel *Empire Star*, for instance, was an "illustration" of itself, even prior to its illustrated edition. In a paradox story, where the reconstitution of *sjuzhet* on the level of *fabula* is in principle unfeasible, this tendency toward visual illustration becomes more pressing than in classical stories, and the paratext therefore tends to foreground itself as a primal means of narrative synthesis and conservation. In essence, the time travel story pushes toward self-depiction, a graphic representation of its own plot, seen from a viewpoint-over-histories that increasingly verges on the physical perspective of a reader holding an actual book, or of a television or film audience sitting before an actual screen.

I have no more faith than Genette, in his brief dismissal of the topic of illustration late in his book *Paratexts*, that I could do any justice to the immense fields covering the question of how human knowledge or stories take visual form. I will therefore propose, very briefly, just two relatively humble examples of the literal diagrammatic construction of plots, each of which could qualify as, or stand in the stead of, a last time travel story. The examples are disparate, which at least suggests how the diagramming of fictional time as a means of "travel" through multiplex viewpoints-over-histories might simultaneously fulfill and overstep the generic demands of science fiction time travel. The first is Laurence Sterne's *Tristram Shandy* (1759–67), specifically volume VI, chapter 49. From the start of the novel, Sterne's narration is famously chaotic: "The action is continually interrupted, the author repeatedly goes backward or leaps forward."[16] Victor Shklovsky, like virtually all formal or structural narratologists following him, observes that "such time shifts occur often enough in the poetics of a novel"[17]—indeed, jumping around like this is what novelistic *sjuzhets* do, even as their conventional deference to *fabular* apriority displaces such shifts into the background of the reader's "naturalized" experience.[18] Thus the graphics that Sterne elects to show us in volume VI, chapter 49 are both exemplary of novelistic structuring generally and radical in the explicitness with which they confess the novel's thwarting of the postulate of *fabular* apriority.[19] It is not simply that *sjuzhet* in *Tristram Shandy* continually resists relegation to *fabula*, since doing that, after all, is just part and parcel of the novel's characteristic metanarration or strategic dissimulation. Rather, *Tristram Shandy* is a time travel novel in its construction of an explicit new viewpoint, one still latent in classical narrative, by which its own irresolution of *fabula* is stridently exhibited as overt paratext. In essence, the novel quite literally becomes a picture of its own narrative lines, and those lines in turn become a physical object in the reader's spacetime, an actual sketch of what impedes his or her power to read (back into) the *fabula*.

Yet Sterne's most bravura gesture of paratextual time travel may be the direct physical insertion, in the middle of the third volume, of a single sheet of marbled paper, the kind used to line the inside of book bindings. Instructing the reader to consider this interposed endpaper as a figure for the work's resistance to elucidation, Sterne writes:

> Read, read, read, read, my unlearned reader! read,—or by the knowledge of the great Saint *Paraleipomenon*—I tell you before-hand, you had better throw down the book at once; for without *much reading*, by which your reverence knows, I mean *much knowledge*, you will no more be able to penetrate the moral of the next marbled page (motly emblem of my work!) than the world with its sagacity has been able to unravel the many opinions, transactions, and truths which lie mystically hidden under the dark veil of the black one.[20]

As what Garrett Stewart terms "demediation," or as what I would suggest is a physical revolt against the postulate of *fabular* apriority, Sterne's paratextual coup is possibly unequaled even in twentieth- and twenty-first-century book design.[21] At such a moment—which for a reader might be fully (or rather fantasmatically) experienced only by possessing the fetish of an intact eighteenth-century edition containing the original marbled page—the novel becomes indeterminately a present (a *sjuzhet*), a past (a *fabula*), a sublimated or demediated nonpresence (a *text*), and an all-too-present, demediated *thing* (a *paratext*). "Read" it, "unravel" it, "penetrate" it, or just "throw down the book at once," as Sterne playfully suggests—the paradoxical structuring of the novel's relation to *fabula* leaves us to our own devices, in a present both freed from and stuck to its specific moment. More precisely, Sterne leaves us with a *leaving*, something like a material memento of our "visit" to another time, like the embroidered handkerchief that a time traveler brings back from an encounter with John Wilkes Booth in the *Twilight*

[152]

C H A P. XL.

I Am now beginning to get fairly into
my work; and by the help of a
vegitable diet, with a few of the cold
feeds,. I make no doubt but I fhall be
able to go on with my uncle *Toby*'s ftory,
and my own, in a tolerable ftraight line.
Now,

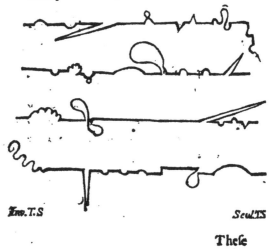

Inv.T.S   *Scul.TS*

Thefe

Tristram Shandy, plot lines

Zone episode "Back There," or the small flowers that Wells's Time Traveller retains from his tryst with Weena.[22]

Tristram Shandy is obviously rather early for a last time travel story; a
somewhat more contemporary work that similarly connects *fabula* to paratext as the means of explicating narratological travel is Alfred Bester's *The
Stars My Destination* (1956).[23] Bester's novel initially focuses on the ability
of characters to "jaunte," a form of mentally controlled teleportation over
limited distances. But at a certain moment of extreme suffering, the pro-

[153].

Thefe were the four lines I moved in through my firft, fecond, third, and fourth volumes.———In the fifth volume I have been very good,———the precife line I have defcribed in it being this :

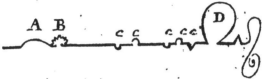

By which it appears, that except at the curve, marked A. where I took a trip to *Navarre*,———and the indented curve B. which is the fhort airing when I was there with the Lady *Bauffiere* and her page,—I have not taken the leaft frifk of a digreffion, till *John de la Caffe*'s devils led me the round you fee marked D.—for as for *c c c c c* they are nothing but parenthefes, and the common *ins* and *outs* incident to the lives of the greateft minifters of ftate; and when compared

pared

tagonist, Gully Foyle, is suddenly goaded to vastly expand his jaunting ability over both space and time, acquiring the ability to revisit and manipulate key moments of his own life history. These self-revisitations had already been anticipated earlier in the novel by vague apparitions of Gully's future self, a "burning man" who intrudes on several scenes, albeit with minimal exposition. In this respect, the progress of *The Stars My Destination*, like that of "All You Zombies—," or indeed that of *Beloved*, entails a circular or self-perpetuating plot, in which, as Žižek argues, a constitutive failure to recognize the full significance of the doubled self or the re-presented past is the motive force that impels the narrative toward its completion.

However, more than the question of self-recognition, Bester is deeply concerned with what happens to narrative viewpoint when the crisis of the *fabula/sjuzhet* relationship is finally brought to a head. As the traumatic incident that gives rise to Gully Foyle's mysterious time-traveling avatars finally arrives, his own spacetime perspective is exploded: "He went hurtling along the geodesical space lines of the curving universe at the speed of thought, far exceeding that of light. . . . [H]is time axis was twisted from the vertical line drawn from the Past through Now to the Future."[24] Foyle's very senses are distorted and jumbled in the hyperspacetime domain of this new viewpoint, and he begins to experience synaesthesia: "Sound came as

The Stars My Destination, Gully Foyle's synaesthesia

sight to him, as light in strange patterns. . . . Motion came as sound to him. . . . Color was pain to him. . . . Touch was taste to him."[25] The ultimate result of this expansion of spatiotemporal consciousness is *graphic*—the narrator depicts Foyle's experience with new spacings and fonts. Finally, then, the *sjuzhet* intrudes directly on the paratext—or rather the paratext directly prints the *sjuzhet*—and Gully Foyle's perspective is rendered in concrete layouts and typescripts, the raw material of publishing. At least as much as in the diagrams of *Tristram Shandy*, transtemporal fiction here becomes literally a published object, one that can make no sense other than as a series of graphic marks on a physical page. Thus the unreconstructable time travel *fabula*, intensely recomplicated by Bester with mythical, psychological, and mystical overtones, nonetheless achieves its climactic moment as a physical *thing*, in the guise of extravagant typefaces, anomalous indents, and hand lettering.

Viewpoint and Multiverse

Contemporary time travel fiction continues to affirm its links to theoretical physics, especially multiverse models and theories. The notion of an ensemble of parallel universes is of course at least as old as the popular reception of Everett's many-worlds interpretation of quantum mechanics in the late 1950s. In science fiction, most travel between parallel universes is enabled by literal technologies, such as Philip Jose Farmer's "gates" between "tiered worlds" (1965), Heinlein's "continua" device for moving between parallel universes (1980), Frederik Pohl's interuniversal "portals" (1986), the remote-control-triggered "wormholes" of the 1980s television series *Sliders*, or even Jane Lindskold and Roger Zelazny's "resonance tracers" and "bottled time" in their computer game *Chronomaster* (1995).[26] In light of my own concerns with how multiple universes are narrated or viewed (in addition to being visited), a last multiuniverse fiction might be Neal Stephenson's *Anathem*, in which "worldtracks" or "cosmi" are literally narrated into being by characters called "Incantors" and "Rhetors," or Greg Bear's *City at the End of Time*, in which "fate shifters" manipulate the interrelations of innumerable cosmological lines. Recent popular films also consider the connection between the depiction of multiplied lines and the problem of narrative viewpoint.[27] In *Minority Report* (2002), *Déjà Vu* (2006), *Source Code* (2011), and *The Adjustment Bureau* (2011), the technology of viewing is explicitly thematized by the film, becoming a kind of literal narratological device. In fact, any one of these films—and there will undoubtedly soon be more—could be considered a last time travel story in the mode of multiplex narratological self-rendering, via its realist depiction of the problematics of surveillance and viewing technology.

In Chapters 5 and 6, following my interpretation of the paratextual structuring of Delany's *Empire Star*, I offered some readings of visual time travel scenarios in *Star Trek* and *Back to the Future*. Initially showing that a multiplex viewpoint-over-histories, while structurally required by narrative generally, is depicted specifically within time travel stories as a literal *position*, I then argued that the time travel film offers something like a full illustration of the subjective conditions of viewing, in which all such requisite viewing positions are already directly portrayed. It is therefore possible to describe the time travel film as a "subject" itself, uncannily abiding in the very place of the subjectivity of its viewer. The counterpart to such a claim is that the human subject is a species of time travel film, a "device" that assembles a material viewpoint-over-histories in order fantasmatically to synthesize, in the *sjuzhet* of its continually renarrated self-identity, multiple temporal lines of *fabula*. Thus time travel, as a basic condition of storytelling—as viewpoint-over-histories, and as paratextual self-illustration—preexists both the literary and the cinematic genres of time travel fiction per se. In turn, the time travel film becomes a privileged site for illustrating the narrative structuring of subjectivity itself. And what the time travel film shows is the infinite scope or expandability of viewpoint, its instrinsic fantasy of wholly free-willed movement through hyperspace-time: the absolute ease by which—perhaps in any *sjuzhet*, tacitly, but very explicitly in a time travel *sjuzhet*—the perspective of narration is cut loose from *all* limitations or concretions of spacetime. As with a film camera, nothing either physical or logical constrains such a viewpoint in principle; it is free to roam over any depicted differences of time and space.

Consequently, my penultimate pitch for a last time travel story is also a film, or rather a filmic type—nonfictional, as it turns out—in which time travel achieves a fully literalized picture of such a universal or metauniversal viewpoint. In 1968, Charles and Ray Eames produced a well-known short film entitled *Powers of Ten: A Film Dealing with the Relative Size of Things in the Universe and the Effect of Adding Another Zero,* based on a 1957 picture book by Kees Boeke, *Cosmic View: The Universe in Forty Jumps.* The Eameses' film opens with the camera looking straight down on a couple picnicking on Chicago's lakefront, a perspective that the narrative voiceover explicitly measures for us: "We begin with a scene one meter wide, which we view from just one meter away." The camera then begins to zoom out as the narrator explains that "every ten seconds we will look from ten times farther away"—100 meters wide, 1,000 meters, 10,000 meters, and so on. The result of this logarithmically expanding zoom is that, somewhere between about seventy and ninety seconds into the film, the shot reaches the limit of any perspective actually achievable via then-current space travel, and planet Earth itself diminishes to a mere point.

It is here that the film starts to become most interesting in terms of viewpoint construction, since now we begin to "observe" the cosmos from perspectives ever more detached from physical possibility: we see, in quick succession, seemingly photographic views of the Earth's orbit around the sun, the solar system, the Milky Way galaxy, the nearby galactic clusters, and finally the entire known universe. At a scale of 100,000,000 light years, the rate of zoom finally slows, and the narrator's voiceover states, "As we approach the limit of our vision, we pause to start back home." By now, of course, we are involved in a time travel story, since the farthest objects observable in an Einsteinian–Minkowskian universe are also the oldest; we have essentially just traveled back 100,000,000 years in time in about four minutes of *sjuzhet* in order to see them. Indeed, an earlier "Rough Sketch" of the Eameses' film includes graphical gauges on the side of the screen—presumably not unlike those used by the Time Traveller in Wells's *The Time Machine*—to measure the difference between the severely dilated "traveler's time" (of the viewer) and the relatively speedy "earth time" (back home). In the film's final version, most of these somewhat busy graphics are removed, and the viewer's travel from frame to frame is marked only by numerical labels indicating the scaled distance from the origin at Chicago's lakeside. What is crucial to note is the absolute ease with which we transition from physical to unphysical viewpoints, and how equivalently conceivable they remain, up to and well beyond the real limit of our possible vision. It seems as if we might zoom in or out forever, in the *sjuzhet*, regardless of

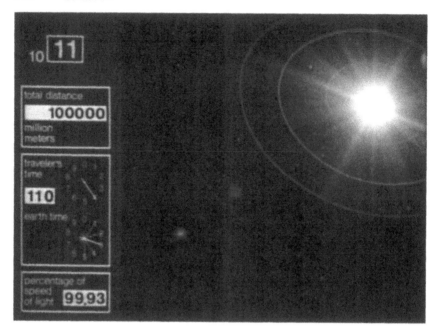

Powers of Ten ("Rough Sketch"), frame with gauges

whatever physical constraints current theory might impose upon any *fabula* of cosmic exploration. Moreover, the initial scene of the Chicago lakefront, viewed after about twenty or thirty seconds as though from an airplane or satellite, is precisely as easy to absorb as the pseudoscene of, for instance, the solar system or the whole galaxy. Throughout, we remain resolutely in the perspectival center of the shot, at the same paratextual distance from our image or object, and within the same-sized frame.[28] Thus the film's ambition to illustrate the immense scale of the universe probably backfires, not exactly because of any failure of its specific technique, but rather by virtue of the general adaptability of narrative perspective itself, its extreme capacity to roam, at effectively infinite speed and yet within a single, steadfast framework, through *any* range of spatiotemporal viewpoints.

I have previously remarked on the capacity of time travel fiction to depict whole universes from otherwise classical narrative perspectives, narrating from simultaneously unphysical and utterly mundane hyper-spacetimes: Jack Williamson's "featureless blue chasm," Fritz Leiber's "Big

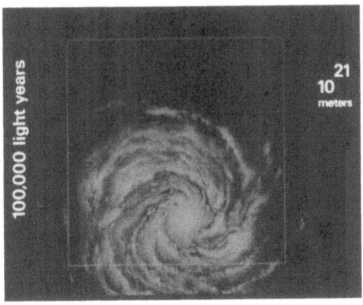

Powers of Ten, two frames

Time," Isaac Asimov's "Eternity," or Rudy Rucker's "superspace." Ultimately, from such positions, the entire universe (or multiverse) can look to be about the same size as this or that planet, or the picnic blanket in Chicago, or, later in the Eameses' film, a molecule, proton, or quark. Indeed, paratextually all these things *are* the same size, as the Eameses' film continually, possibly unwittingly, reminds us by framing its ten-second intervals of zoom with a series of receding graphic boxes, one new box for each power of ten. The logarithmic scale of expansion or contraction ironically abets the radical domestication of scale, since each order of magnitude is reduced to an essentially linear representation. What the film's *sjuzhet* constructs is therefore decisively not the *fabula* of a journey to the outermost or innermost limits of vision, but instead a fantastically reductive paratext, a single *thing* that unremittingly remains the *same* size, always immediately before us here and now.

As a result of this extreme, even hyperbolic reduction of *fabula* to visual paratext, the entire universe may be rendered as an object at hand, much like the tiny ball of wax Descartes holds in front of him as he sits by his fire in Western philosophy's most famous narrative of armchair speculation.[29] The outer limits of narrative adventure are utterly conquered, enclosed by the framework of a single screen—a "universal picture," to paraphrase Heidegger's well-known critique of the reductionism of modern technology. *Powers of Ten* is a paragon of time travel's tendency toward conservation, the reduction of potentially radical plot difference to synthetic completion and closure.

Incidentally, in numerous recent nonfiction films about scale indebted to the Eameses, for instance in the American Museum of Natural History's 2009 production *The Known Universe*, obvious attempts are made to decenter the viewer's intensely focused or confined perspective, as well as to update the cosmology. Certain advantages are presumably gained over the Eameses' film in the increased scale of *The Known Universe*—14 billion light years instead of 100 million—as well as in the less stubbornly human-centered camera shot.[30] However, if anything, the increase in scope and distance reinforces the impression of a universe assimilable as a single visual icon, portrayed for five or ten minutes onscreen—a universe, in a word, *consumable*. In terms purely of graphical content, and despite the updates in underlying physical theory, our viewpoint is not really different

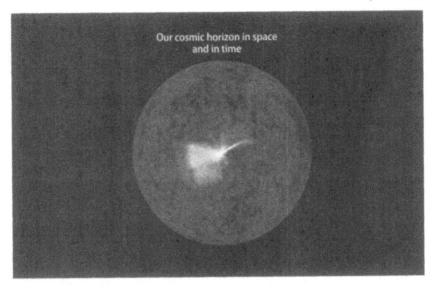

The Known Universe, "Our Cosmic Horizon in Space and Time"

The Ptolemaic universe

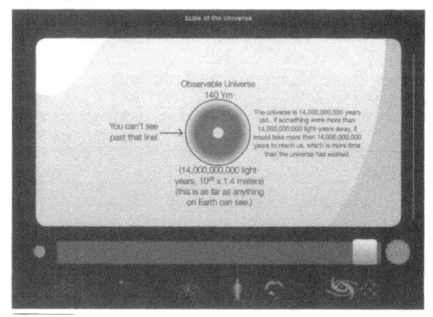

"Scale of the Universe," "You Can't See Past That Line!"

from that of the Renaissance reader observing a diagram of "The Ptolemaic Universe." Film technology and a quasi-theological infinitude, something like a god's-eye view, are rendered effortlessly compatible.

I wish to mention one more version of the nonfiction "scale" film or diagram that perhaps exemplifies this dual tendency to expand and contract the universal image, a finely conceived and produced interactive internet tool by Cary and Michael Huang entitled "Scale of the Universe." In lieu of merely watching viewpoints get wider or narrower, as in *Powers of Ten*, the web page user manipulates a slider to increase or decrease the scale of the frame at will. Arriving at the scale of the "Observable Universe," he or she encounters a circle labeled "14,000,000,000 light years, $10^{26} \times 1.4$ meters." An arrow points to the circle, ironically captioned, "You can't see past that line!" True, in the *fabula*; yet of course one *can* see past the line. One can even use the slider to zoom farther out, not only to the "93,000,000,000 light years" of the "Estimated Size of the Universe" in current inflationary theory, but beyond even that—to what? Well, to the paratextual frame, centered neatly on the computer screen, the purely formal limit posited by the last position the pseudophysical slider of the paratext is programmed to allow.[31] Outside of everything, there is still something—the hyperspacetime of narrative viewpoint, physically rendered as a mouse movement traversing the entire universe, a magnificent, mundane transcendence.

Science Fiction Redux

Returning to the most familiar genres of time travel literature and film, we might select a last time travel story from among a rich array of recent fictions and films that play self-consciously or parodically with their own temporal mechanisms of story construction—this is the criterion I have generally preferred to distinguish time travel, rather than specific plot contents or technologies: *Inception, Primer, Timecrimes, The Fountain*, several of the *Harry Potter* films, and so on. But the purest version of such a time travel story may be, ironically, 2010's *Hot Tub Time Machine*, a film that, like *Back to the Future* (which it references extensively), is far more sophisticated regarding the generic, sociocultural, and institutional elements of its own production than one might be led to suspect by its own marketing.

In nearly every aspect, *Hot Tub Time Machine* is an unabashedly derivative film, almost entirely a pastiche of prior time travel stories and motifs, borrowing from works such as *The Terminator, Peggy Sue Got Married, Eternal Sunshine of the Spotless Mind, The Butterfly Effect*, and episodes of *The Twilight Zone* and *The Outer Limits*. Three friends visit a ski resort, intending to re-create for themselves a memorable weekend they spent together some twenty-five years before. Accompanying them is one of the protagonist's nephews, who had not yet been born at the time of their first visit. Early in this outing, the group suddenly find themselves transported back to that prior weekend by a malfunctioning time machine unaccountably built into the hot tub at their hotel. Back in the 1980s, entirely typically and (therefore) comically, they set about confronting the causal subplots that gave rise to their future selves, and that may now (possibly) be amended.

The plot of *Hot Tub Time Machine*, like that of *Back to the Future*, is a type I have called "conservative," in which the future to which the protagonists eventually return is an improved variant, constructed in terms suited both to their own sentimental cravings and to the generic constraints of the middle-aged buddy film. As in Robert Silverberg's *Up the Line*, my first example in this book, one of the characters, Lou, remains behind in the past to carry out the temporal labor necessary to create this improved future, while the other characters jump instantaneously ahead to discover their lives revised, yet with minimal uncanny embarrassment—they are now wealthy, happily married, and professionally successful, and have acquired whole new pasts to undergird these altered personae. The absurd over-drawing of the historical revisionism accomplished—Lou will have founded Google (now called Lougle) and will have been the lead singer of Mötley Crüe (Mötley Lüe)—is entirely in keeping with the hyperbolic parody that time travel writers have very often (arguably always, sometimes despite themselves) engaged in, from Mackaye to Heinlein and beyond. *Hot Tub Time Machine* pushes such implausible renovation to a comic extreme enabled by its own self-conscious or sardonic pop-cultural capital, blithely disregarding both dramatic consistency and psychological implication. The loss of twenty-five years of identity by the characters returning to their altered futures is, like Marty's loss of his original family in *Back to the Future*, entirely elided by the ridiculous but generically consistent tautology that the "better" future is preferable. *Hot Tub Time Machine* is thus an unparalleled extrapolation of the perhaps inherent fantasy of a happy ending that generic time travel—in collusion with popular culture in general—erects against the terrible inexorability of the factical. Moreover, the movie's postmodern nonchalance, its carefree attitude toward the possible social, sociological, psychological, or historiographical consequences of the intratemporal juxtapositions it constructs, allows it to fulfill an aim of time travel fiction inherited from its earliest generic manifestations: the macro-logical (nigh paratextual) depiction of the structure of storytelling itself, along with the suppression of the social and political content actually carried or connoted by cultural history. As a tale about other time travel tales—therefore also as (1) a parody of the strictly generic demands of any textual or paratextual assemblage of narrative lines; (2) a restoration of *fabular* apriority in the mode of a self-consciously paratextual exhibition of pop-cultural imagery; and (3) an ironically realist depiction of the ideological unrealism of its chosen pair of cultural eras—*Hot Tub Time*

Machine is as close as one might come to the last time travel story, for now anyway.

But I have no desire to finish the book with quite so bluntly a paratextual or parodic example. Instead, let me offer one more suggestion for a last time travel story from some years earlier, and for reasons more appropriate to the structural-narratological inquiry I have pursued in lieu of a more strictly chronological cultural history of time travel: Harlan Ellison's 1970 short story "One Life, Furnished in Early Poverty," a text I will analyze alongside its television adaptation in the 1980s *Twilight Zone* series.

Ellison's story is about a middle-aged script writer named Gus Rosenthal, melancholic and bitter in his world of adult business dealings, who feels an irresistible nostalgic urge to revisit the Ohio town in which he grew up. Arriving at his former house and searching about the lawn for toy soldiers he had once buried there, Gus abruptly finds himself transported back thirty-five years, his family, town, and younger self revivified; suddenly he is a potentially intruding witness to his own seven-year-old life. The story dabbles in paradox, initially in the familiar mode I identify as a "conservative" approach, although in this case with more fatalistic or tragic consequences. Gus, adopting the ambiguous pseudonym "Harry Rosenthal," meets and befriends his childhood self and attempts to intercede in what he remembers to be key traumatic moments of his early life: he prevents bullies from harassing the boy, explaining to the boy's father (as he couldn't do when he was young) that Gus did not start these fights; he persuades young Gus to consider career paths that might make him happier as an adult, and so on. However, eventually Gus/Harry feels himself obliged, through a vague urge similar to the one that first brought him back to Ohio, to depart the past and return to his proper time. In preparing to do so, and informing his younger self that he will have to leave, he discovers not only that he has failed to protect Gus from the traumas that lead to a sad and bitter adulthood, but that it is his *own* anachronistic appearance and abrupt departure within the past—the boy's now unrepressed memory of the desertion of "Mr. Rosenthal"—that will have set the child onto the regretted and resentful path.

In the original story, Ellison narrates the young Gus's reaction to Gus/Harry's abandonment with an ambiguous mixing of "I" and "him" that cogently jumbles the adult's first-person voice with the child's inner monologue, as the hazily remembered primal moment of embitterment reveals itself:

> I watched him go away. He was my friend. But he didn't have no guts. He
> didn't. But I'd show him! I'd really show him! I was gonna get out of here, go
> away, be a big person and do a lot of things, and some day I'd run into him
> someplace and see him and he'd come up and shake my hand and I'd spit on
> him. Then I'd beat him up.[32]

It is difficult to ascertain the precise degree to which the boyhood voice is projected or reimagined by the adult viewpoint through which the story continues to be focalized. Ellison's ending insightfully leaves open questions about how much of the young Gus's reaction comprises the *nachträglich* projection of the adult's melancholia, and more generally, about the intensive and complex psychological relationship between childhood memory and adult personality. Here, as in "All You Zombies—," the time travel story literalizes a narrative goal shared by the methodology of psychoanalytic therapy: to revisit, or more precisely to rediscover or even redetermine, the primal scene of adult subjectivity, and then to renarrate the precise extent of that primal scene's effectivity, still scrutinizing the comparative consequences of present and past cathexes. While Gus/Harry comes to the conclusion, with the confident literal-mindedness of the time-traveling character, that his own direct anachronistic intervention— "what I'd done to him, forced him to become"—is the primal trauma in young Gus's life, the story overall projects a less certain decision about the roots of Gus's disaffection, or about the continuity of the narrative that connects it to its prehistory. Ellison's story ends with a direct but strangely cryptic first-person rumination: "But . . . oh, God . . . what if I came back again . . . and again? Suddenly the road did not look familiar."[33]

The *Twilight Zone* adaptation of Ellison's "One Life" seems to oscillate between maintaining such an uncanny and fruitful psychological ambiguity and papering it over. In the screenplay, what had originally been young Gus's interior monologue, the transcript of a perhaps indistinct memory, is now rendered as straightforward dialogue, shouted by the boy angrily at the older Gus/Harry: "Some day I'll run into you, and you'll come up to shake my hand, and then I'll spit on you, I'll spit right in your face, and then I'll beat you up." This outburst is followed by Gus/Harry's concluding soliloquy, which exposits somewhat redundantly the story's apparent loop:

> Oh my God. I remember now. I remember Mr. Rosenthal. I remember when I left me. It wasn't them. It wasn't any of them. It was Mr. Rosenthal. It was me.

Gus/Harry's own effect upon the past, now remembered or unrepressed, had already been part of his *fabula*, and nothing will have changed by virtue of his anachronistic intervention after all.

Regardless of this slightly ham-handed final speech, which doesn't exist in the original story, the episode makes very clear from the start that Gus's resentful aspiration to "be a big person" is his chief fault. By the moment of the climactic scene in past Ohio, it is abundantly obvious that the time traveler is/was the catalyst for that flawed aspiration. However, the overdirectness of the exposition, whatever its aesthetic merit, stands quite apart from the real significance of the time travel story, a significance that emerges, as usual, rather by way of the mechanism of storytelling itself and the literalization of its structuring conditions. The anachronistic advent and subsequent abandonment of "Mr. Rosenthal" in the life-history of the child is rendered visual and empirical, literalizing the process by which the present revisits and inflects the past event: "It was me," not hypothetically, fantasmatically, or allegorically, but *really*, as I now "remember" it in the form of a generic story. "I" am there to *view* the primal continuity of childhood and adulthood, along with the primal origin of "my" (story of) subjectivity, in the (recovered) traumatic moment of its original rupture. As in Heinlein's "All You Zombies—," the reader can observe this complex fantasy as an explicit parody of self-scrutiny, a fascinating overreaching of the possibilities—dryly narratological, in one sense, but potentially horrible in another—of revisiting one's past.

Thus, with the conclusion of "One Life," as with Michael Moorcock's *Behold the Man*, which I discussed at length in the Introduction, we confront the profound intersection of psychology and historiography: the fundamental question of the cathexis of the past event. All issues of the compulsion toward *fabular* apriority in a sense revolve around this intersection, at which the question gets asked: *why* is a particular moment of the past still present in a narrative? It would perhaps not be an exaggeration to say that Ellison's story, especially in the combination of its written and filmed forms, is here opening up—once more, in literal form, as a tale with characters, scenery, and dialogue—the question of the psychological ground of historiography itself, the ineluctable and fertile ambiguity between the past as *causing* and as *caused*, and of the scenography in which that foundational ambiguity is re-viewed.

In noticing the elemental level at which the subject's life gets unfolded at this realistically depicted juncture of Ellison's story, we also find ourselves viewing—perhaps as always in a time travel story—the paratext, the most immediately perceptible arrangement of narrative lines and of any potential philosophical inquiry into their relation and juxtaposition. It should not be surprising, then, that the fullest and most reflective register of psychological and historiographical inquiry opened by "One Life" will occur on the fully paratextual level of the story—within a con-text in which it is no longer possible either to disregard or to bracket the narrative's connections to the material life of its author, reader, producer, and viewer—and to which I now turn.

The 2004 DVD packaging of the 1980s *Twilight Zone* series also contains the usual assortment of special features, including voiceover commentaries by Harlan Ellison produced specifically for this edition. The commentary for "One Life, Furnished in Early Poverty" begins with Ellison in a sound booth (as he lets us know), watching the episode and offering autobiographical background leading up to his family's move to "Painesville, a little town about 30 miles northeast of Cleveland."[34] The autobiographical details are especially pertinent here because, as Ellison asserts, "there's more of my actual life in this than in any other story I've written." Furthermore, the background for "One Life," even amid the affable sentimentality of Ellison's recollections—he begins by singing the blues standard "After You've Gone," before introducing himself to the audience—is far from idyllic:

> At the time, we were the only Jewish family in the town. And while Ohio flies under the banner of being the great American heartland, and filled with wonderful people, and it has produced some wonderful people like Milton Caniff, and the woman who wrote *My Sister Eileen*, and Gardner Rea, the great New Yorker cartoonist, and a bunch of other impressive people, it is also a place where, in pre-1940s and probably very late after that, it was the heartland full of antisemitism, bigotry, racism, and the worst kind of male-chauvinist misogynism you can imagine.

Living in this benighted place, and also being a relatively "little kid," the young Ellison is mercilessly bullied: "[T]hey used to beat the crap out of me every day in the schoolyard." Ellison explicitly characterizes these attacks as provoked by an incoherent but still vicious antisemitism, alongside more mundane juvenile malice.[35] Some part of the adult Gus/Harry's intervention, as Ellison suggests while watching his story redepicted, is therefore a nostalgic redress for this early victimization: "[W]atch this scene," Ellison tells us, "watch me running away from the gang. I never had a chance to run away most of the time, because there was an enclosed school yard. But I ran away, and there I am, saving me from the bullies."

However, the main portion of Ellison's commentary concerns his childhood relationship with his parents, indeed quite a bit more than one might expect either reading the original story or watching the filmed version. After one particular episode of bullying, in which the young Ellison is left bruised and naked with his clothes ripped off, he hides, mortified, in the cold. In the voiceover, Ellison sets up the ensuing encounter with his parents by telling us, "I was ashamed to go home, so I hid in a bush, the snow, for hours and hours and hours, and three, four hours later, just blue, I went home."[36] When Ellison finally does return home, the following conversation takes place:

> I came in, and my mother was frantic, and after they had done the usual thing
> of, you know, where were you, and don't you ever do that again, and, you know,
> all of that, and hugging me because they're so relieved, and doing all of that,
> and my mother said, what had happened, and I told her, and she said something
> to me that stuck with me. She didn't mean to be cruel, and she didn't mean
> to be . . .

At this point, literally in midsentence, Ellison breaks off, drifting into a digression about the toy soldiers he used to bury in his yard, a framing motif in both the original story and the television version. For several minutes, Ellison declines or fails to return to the topic of his mother's "cruel" assertion, and instead extemporizes about these soldiers, then about the actor Peter Riegert, who plays the adult Gus/Harry in the episode, and finally about some thoroughly nostalgic memories of his father—"Dad, you'd be so proud of me. I turned out so well." Finally he returns to that tantalizingly interrupted speech of his mother:

Anyway, where was I? I get, uh, I get the willies when I talk about this script
and this story. They beat me up and my mom said to me, what did you say to
get them angry?

This disturbingly callous and oblivious question from Ellison's mother
seems to me to be the true core of the *fabula*, the moment of his history
when the trauma of childhood and the compulsion to renarrate in the form
of time travel, to redepict the primal origin of adult history and subjectivity,
takes place. Or rather, productively, ambivalently, the advent of the trauma
fails to take place: where the mother's compassionate acknowledgment and
response should be, instead there is a horrible gap. It is probably impossible to track down the missing nucleus of Ellison's mother's disturbing
question: is it mere naïveté, or is it a phobic refusal to confront the antise-
mitic subtext of the attack on her son, or, finally, is it an entirely general
displacement of the "antisemitism, bigotry, racism, and . . . misogynism"
that Ellison himself recalls, and which must have inflected the family's—
and particularly the mother's—entire quotidian life in mid-twentieth-
century Painesville, Ohio? Whatever the (presumably) multifarious ground
of his mother's errant question, Ellison's apparently still raw reaction to its
memory indicates a failing of parental empathy quite different from, for
instance, that of the father, whose misinterpretation of Gus's fighting is
retroactively corrected by the time-traveling Gus/Harry. No, the mother,
far from being muddled or offering misplaced love, instead reemerges in
Ellison's grudging reminiscence only as dreadfully, irrecoverably *indifferent*: the very dialectic of a correctible misunderstanding seems missing
from her speech.

Such a moment is perfectly narratable, of course, but its full cathexis
might be indicated only by ellipses and lacunae, in this case the uneasy and
truly poignant halt of Ellison's own recollection in the commentary, and its
temporary redirection through the more familiar *topoi* of the time travel
plot and the sentimental memories of his father. Let me propose the following interpretation of the multilayered relationship between Ellison's
story, its cinematic adaptation in *The Twilight Zone*, and the DVD commentary that finally gets appended to it. The primal moment occurs not in the
story itself, nor even in the narrative reencapsulation that Ellison's commentary provides, but rather in the *hiatus* of that narration, in the stum-

bling of its very syntax, and in the temporal breach opened up by the pain of remembering and having to retell. The fully paratextual commentary thus contains—and embodies, in the ambivalent form of its own *failure* to have been represented directly in the "primary" narrative of either the short story or the television episode—the *first* moment of the fiction, its true origin, now projected back. Ellison "get[s] the willies" at this moment, no doubt. So he proceeds to improvise a possible response to his mother's speech, albeit more in the mode of exasperated sarcasm than of nostalgic recuperation:

> [briefly laughs] If I had been a little older and a little less respectful I would have said, yeah, Mom, sure, you know, I just yanked their chains so much that six of them beat the crap out of me. But I didn't say that. I just started to cry, because I'd embarrassed my mother and my father, made them feel ashamed of me. Well [sighs], that's part of the story.

Of course, this response really is *not* "part of the story"; the retroactive explanation Ellison offers to his mother in the commentary is precisely not what we observe in either the text or the adaptation. Or rather, it is "part of the story" only as a primal psychological—and perhaps sociopolitical—moment sacrificed to the textual exigencies of time travel genre and the plot devices it entails, at most partially or obscurely observable through a supplementary, paratextual revisitation.

This is time travel: the halting and ambivalent return, from the past, of the primally cathected event, uneasily incorporated into an unfolding future that dissimulates itself in the form of generic narrative. In fact, everything *objective* in Ellison's story—the toy soldiers that Gus/Harlan feels compelled to return to dig up, the fantasmatic restoration of Gus/Harlan's relationship with his father through time travel, the whole drama of bullying and childhood insecurity—revolves uneasily around the deep diachronic lacuna opened up after the fact in Ellison's commentary around that terrible, even unintentionally brutal, indifference of the mother's question. But therefore the full beauty of the story, and of its horror, is accessible only through a retrospective juxtaposition of its various paratextual strata: the literal plot, inflected with both explicit and tacit autobiography; the television episode, in which the autobiographical tableau of return is visually fulfilled; and the supplementary commentary, which explicitly reconnects the story's temporal strands to its autobiographical sources—"I'm watching me; Peter Riegert is playing *me*, as an adult, returning to my childhood!"[37] Finally, only then, might we arrive at the psychological or sociopolitical core of the story, in which words that should have been spoken in the past simply weren't, or couldn't be—a failure that simultaneously compels the return of the time traveler in the guise of realist plot and ultimately renders that return an unrealizable fantasy.

1. Silverberg, *Up the Line*, 186–87.

2. Of course, I am momentarily ignoring the story's other extraordinary event, Jud's liaison with his own ancestor. I will revisit the significance of such oedipal or quasi-oedipal encounters especially in Chapter 6.

3. Bal, 217.

4. Genette, *Narrative Discourse*, 35.

5. Jonathan Culler offers a useful survey of the varieties of this distinction in structuralist theory and narratology, as well as of its basis in Russian formalism (see *Pursuit of Signs*, 169–87).

6. Genette, *Narrative Discourse*, 35–47; see also Bal, 82–88.

7. Genette describes the sacrosanct position of the *fabula* in "classical" narrative, a type that he considers opposed to post-Proustian modern narrative, as "domination" by *fabula*. Speaking of instances in Proust in which "the time of the story . . . regains its hold over the narrative," Genette writes, "It is in fact as if the narrative, caught between what it tells (the story) and what tells it (the narrating, led here by memory), had no choice except domination by the former (classical narrative) or domination by the latter (modern narrative, inaugurated with Proust)" (*Narrative Discourse*, 156, 156n).

8. Note that John Brunner suggests a similar theme in his novel *Times Without Number*: "[T]here were three zones of history which had exercised an obsessive fascination on temporal explorers ever since the Society was founded. One, inevitably, was the beginning of the Christian era . . . but access to Palestine of that day was severely restricted for fear that even the presence of nonintervening observers should draw the attention of the Roman authorities to the remarkable interest being generated by an unknown holy man, and cause Pilate to act earlier than the Sanhedrin, according to the written record, had desired" (145, Brunner's ellipsis).

9. Moorcock, *Behold the Man*, 26–27.

10. Ibid., 49.

11. White, "Question of Narrative," 29. White is partly quoting Hegel's *Lectures on the Philosophy of History*.

12. Moorcock, *Behold the Man*, 135.

13. Ibid., 145–46.

14. In an earlier book, I discuss a similar historiographical problem in the context of the formation of philosophical canons and histories, and argue that critics and philosophers construct, through revisionist rereading techniques, histories of philosophy that must be seen as simultaneously retroactive and proleptic, essentially texts with "no present." See Wittenberg, *Philosophy, Revision, Critique*, especially Chapters 3 and 4 and the "Interlude."

15. Niven, "All the Myriad Ways," 6–7. Unless otherwise noted, I cite the version of the story in Niven's collection *All the Myriad Ways*.

16. See Everett, "Theory of the Universal Wave Function" and "'Relative State' Formulation."

17. DeWitt, 161.

18. Niven, "All the Myriad Ways," 7.

19. For an excellent discussion of incompleteness and fictionality, see Ronen, 114–43. Also see Doležel, who usefully cites a number of sources in analytic philosophy for this definition of fictionality (22f).

20. Niven, "All the Myriad Ways," 7 (Niven's ellipsis and punctuation).

21. Niven, "Preface," 70.

22. Ibid.

23. Nietzsche, 194.

24. Niven, "Preface," 70.

25. Niven writes elsewhere that "a writer who puts severe limits on his time machine is generally limiting its ability to change the past in order to make his story less incredible" ("Theory and Practice of Time Travel," 111).

26. Niven, "All the Myriad Ways," 7.

27. This question of simulacra is not entirely a common motif in time travel stories, but one that is occasionally invoked both in highly philosophical instances and in highly popular ones—or in some cases both. In the popular vein, compare Michael Crichton's *Timeline*:

> "The person didn't come from our universe," Gordon said. . . .
> "So she's almost Kate? Sort of Kate? Semi-Kate?"
> "No, she's Kate. As far as we have been able to tell with our testing, she is absolutely identical to our Kate. Because our universe and their universe are almost identical." (180)

28. Dick, *A Scanner Darkly*, 100.

29. Niven, "All the Myriad Ways," 7.

30. This problem is cleverly elaborated in a 1994 episode of *The Simpsons* entitled "Time and Punishment": Homer, upon returning in a time machine to a present that clearly diverges from the one he left—for instance, his family members now all have forked tongues—exclaims, "Ah, close enough."

31. Laumer, 51.

32. Niven, "All the Myriad Ways," 7.

33. Todorov discusses the difficulty of establishing any such "smallest narrative unit" (48f). See also Dorfman, passim.

34. Niven, "All the Myriad Ways," 10.

35. Ibid. (all ellipses Niven's).

36. Ibid., 11 (Niven's ellipsis). In the reprinted version of the story in Niven's collection *N-Space*, the concrete arrangement of these lines on the page is altered:

> And picked the gun off the newspapers, looked at it for a long moment, then dropped it in the drawer. His hands began to shake. On a world line very close to this one . . .
> And he picked the gun off the newspapers, put it to his head
> and
> fired. The hammer fell on an empty chamber.
> fired. The gun jerked up and blasted a hole in the ceiling.
> fired.
> The bullet tore a furrow in his scalp.
> took off the top of his head.
> "All the Myriad Ways" [*N-Space* version], 80 (Niven's ellipsis).

37. See Leibniz, §60–62 (276–77).

38. Wheeler, 151.

39. Ibid.

40. Ibid, 152.

41. Ibid.

42. Wheeler, responding to Everett, states: "The word 'probability' implies the notion of observation from outside with equipment that will be described typically in classical terms. Neither these classical terms, nor observation from outside, nor a priori probability considerations[,] come into the *foundations* of the relative state form of quantum theory" (152).

43. Again in Wheeler's words: "The model has a place for observations only insofar as they take place within the isolated system. The theory of observation becomes a special case of the theory of correlations between subsystems" (151).

44. There may be interpretations in which the inhabitation of a metaworld is not "unphysical," albeit in precise and limited senses. In Chapters 3 through

5, I allude at greater length to the physical model of a metaverse, which in certain aspects parallels the metaworld imputed within the metaphysics of Leibniz's *Monadology*. In "All the Myriad Ways," Niven calls it a "megauniverse of universes" (1).

45. For instance, see Calhoun-French on contemporary romance time travel; see Radway for a more general discussion of the significance of the romance genre.

46. Paul J. Nahin exhibits such a bias in his book on time travel, which, despite its admirable expansiveness, begins by severely limiting the possible domain of the study of time travel fiction: "In this book we are interested in physical time travel *by machines* that manipulate matter and energy in a *finite* region of space. . . . In addition, the machine must have a *rational explanation*" (18, Nahin's emphasis). Thus Nahin excludes, for instance, any story that uses "mind travel," "dreams," "drugs," "freezing and sleeping [into the future]," "channeling," "accidents" (as in Twain's *Connecticut Yankee*), "illness" (as in Octavia Butler's *Kindred*), "time portals that look like green fog," "psi powers," any technology lacking an "explanation" or a basis in "rationality and science," and so on. Needless to say, such criteria preclude a great many narratives, time travel or otherwise, from critical scrutiny. Nahin's bias, a quite common one in the context of "hard" science fiction, emerges most clearly in his tendency to equate "interest" with "hardness": "More interesting—that is, more rational . . ." (16); ". . . interesting to us because they are rational" (21); "Interesting, yes, but it isn't physics . . ." (16), and so on. The canonical "hard" science fiction writer Arthur C. Clarke makes the point in a similar way, listing and judging the following catalog of "scientific" possibilities for fiction: "immortality, invisibility, time travel, thought transference, levitation, creation of life. For my part, there is only one of these that I feel certain (well, practically certain!) to be impossible, and that is time travel" (173). Gregory Benford, an equally "hard" author who *does* write time travel, claims that "a science fiction writer is—or should be—constrained by what is, or logically might be" ("Exposures," 247). I discuss the applicability and the limitations of a logical constraint on time travel narratives in Chapters 4 and 5. For useful historical discussions of the advent of "hard" science fiction, see Luckhurst, 66–91, 99ff; and Bould and Vint, 74–91.

47. Silverberg, "Introduction," x.

48. See Hayles, especially 111–37; Lem, "Time-Travel Story," 26–33; Penley, especially 106–18; Sobchack, *Screening Space*, 223–305; Landon, *Aesthetics of Ambivalence*, especially 74–83; G. Stewart, *Framed Time*, 122–63. I can mention in passing that there are, perhaps surprisingly, also a fairly large number of works dealing with supposedly actual time travel, either as something already accomplished or as soon forthcoming—usually with little basis in either science or literary theory, although sometimes with ties to various pseudosciences

and to New Age metaphysics. For the most part, I deliberately decline to distinguish subtypes of the genre of science fiction here, for instance "hard" science fiction, new wave, cyberpunk, and so on, with the understanding that the genre of time travel fiction evolves relatively independently of these types, and to a great degree in an uneasy tension with the larger sociocultural history of science fiction. Some sense of this tension may be gleaned by noting the relative absence of discussions of time travel fiction as a distinct story type within excellent critical studies of science fiction such as Bould and Vint's *Concise History of Science Fiction*, Luckhurst's *Science Fiction*, or Csicsery-Ronay Jr.'s *The Seven Beauties of Science Fiction*. These critics mention time travel, for instance, as a "framework" for addressing other themes or sociopolitical issues (Luckhurst, 195; Luckhurst's approach is relatively consistent with Suvin's foundational understanding of science-fictional structuring as "cognitive estrangement"), or as a narrative means to permit "a conscious mortal being [to] return to its origins" or to imagine "the permeability of history" (Csicsery-Ronay Jr., *Seven Beauties*, 99, 100). Such understandings of the usefulness of time travel within science fiction are, of course, compatible with my analysis of its functioning as a narratological laboratory, but they also help explain why time travel has tended to avoid falling into a distinct subgeneric niche within science fiction at large, and has instead cut across more clearly visible delineations of aesthetic, stylistic, and political subtypes.

49. See Gerrig, 10–17.

50. This latter problem is Gérard Genette's chief focus in *Narrative Discourse*.

51. During this first period of time travel literature, the single work most often identified as an origin is also written: H. G. Wells's *The Time Machine* (1895). For a number of reasons, which I discuss in the opening chapters, I decline to grant such credit to Wells's book. However, I discuss *The Time Machine* at greater length in my "Historical Interval II."

I. MACROLOGICAL FICTIONS: EVOLUTIONARY UTOPIA
AND TIME TRAVEL (1887–1905)

1. The first use of a time machine is usually credited to H. G. Wells, in his 1888 short story "The Chronic Argonauts," a precursor to *The Time Machine* (1895). David Smith notes that Wells may have been working on versions of this story as early as 1884 (46). A virtually contemporaneous work that might claim the first time machine is Enrique Gaspar's *El anacronópete* (1887); Wells was almost certainly unware of it.

2. Bellamy, 23.

3. Ibid., 24, 33.

4. Ibid., 30.

5. Ibid., 33.

6. Darko Suvin notes that Bellamy's "refus[ing] the easy alibi of it all being a dream—a norm from Mercier and Griffith to Macnie—marks the historical moment when this tradition came of age and changed from defensive to self-confident. The new vision achieves, within the text, a reality equal to that of the author's empirical actuality" (177).

7. Richard Lanham defines macrologia as "long-winded speech; using more words than necessary" (96). Puttenham's more concise definition is "Long Language" (343).

8. For a useful general discussion of the influence of Darwinism in the United States, see Russett's *Darwin in America: The Intellectual Response, 1865–1912*. In American literary history, the theory of evolution is usually cited as an impetus for the turn *from* realism to naturalism, circa the 1890s. Michael Davitt Bell calls this turn, along with its Darwinistic basis, a "virtual axiom of our literary history" (109). Donald Pizer, in terms closer to those I wish to emphasize, characterizes the basis of the turn as a "half-understood Darwinism and a too-readily absorbed Spencerianism" (*Twentieth Century American Naturalism*, 4).

9. Ruse, 229.

10. Despite some distractions from pure science, in particular the Civil War, Darwin's ascension proceeded more rapidly in the United States than in his home country of England; by 1869, he had been made an honorary member of the American Philosophical Society. It is worth noting also that the surname "Darwin" was already widely known and associated with evolutionary theory, both in England and the United States, because of the fame of Charles's father, Erasmus.

11. These theories include Lamarckism, which proposes the inheritance of social traits and behaviors; various forms of "catastrophism," offered largely in response to the geologist Charles Lyell's popular "uniformitarianism"; a number of developmental models influenced by the fossil studies of Cuvier, Owens, and other paleontologists; and the "idealisms" of writers such as Louis Agassiz and Robert Chambers, who, like nearly all of these other theorists, explicate what is routinely assumed to be the "progress" of evolution toward a contemporary human (that is, western European or even Anglo-Saxon) society. For a discussion of evolutionary theory up to 1859, see Bowler, 96–133.

12. For instance, see Samuel Butler's *Evolution, Old and New*, in which the author, amidst an elaborate homage to Darwin, attempts to square "natural selection" with older, teleological models of evolution, especially Lamarck's, but also those of Buffon and Erasmus Darwin. Butler's argument revolves around the apparent fact that Darwin ambiguously offers natural selection as a mechanism, but not a cause, of species variation, and therefore leaves open the possibility of a purposive force in the formation of species. Bowler, citing others' surveys of the periodical presses, notes that "by 1870 the basic idea of

evolution was becoming widely accepted, and because the public associated the idea with Darwin, the name 'Darwinism' became popular," but that "[b]y today's standards, many of [the] early evolutionists were only 'pseudo-Darwinists'—they were unified only by their recognition of Darwin as the figure who had prompted their conversion to the general idea of evolution" (188). Compare Donald Pizer:

> [T]he basic pattern of evolutionary change . . . was seldom Darwinian. Rather, most critics accepted and absorbed Herbert Spencer's doctrine that evolution is, in all phases of life, a progress from the simplicity of incoherent homogeneity to the complexity of coherent heterogeneity. There were several reasons why the Spencerian formula appealed to literary men. It was universally applicable, explicitly optimistic, and easily grasped; it was capable of wide variation depending on the predilections of the individual writer. (*Realism and Naturalism*, 88–89)

Even Darwin himself can be viewed as what Bowler calls a "pseudo-Darwinist," capable, in *The Descent of Man*, of sacrificing the rigor of his earlier work sufficiently to borrow back Spencer's term "survival of the fittest" in order to speculate about the supposed social evolution of the races (136).

13. Richard Hofstadter famously identifies "the United States during the last three decades of the nineteenth and at the beginning of the twentieth century" as "*the* Darwinian country" (4–5).

14. Gerber, 3ff.

15. According to Phillip Wegner, Lew Wallace's *Ben Hur* and Harriett Beecher Stowe's *Uncle Tom's Cabin* were the only two other works of the nineteenth century as widely read and disseminated (63); Brian Stableford asserts that *Looking Backward* "rapidly overhauled" *Uncle Tom's Cabin*, ending second only to the Bible (25).

16. Charles Rooney has a useful discussion of the hegemony of the evolutionary prototype in late-nineteenth-century and early-twentieth-century utopian fiction (27–31).

17. Bellamy, 49.

18. Louis Budd cites Roger Smith as observing that Darwin had "'established a theoretical framework for integrating biological thought with the mechanistic structure of physical thought,' thus supplying 'grounds for a unified system of knowledge'" (28).

19. Chavannes, 105. He continues: "'In the old times of which we were just now speaking, the conditions were such that the highest positions were attained by plundering those who could not defend themselves. . . . [W]hen society improved its organization so that open violence ceased to be profitable and the resistance to acquisition by force had sufficiently increased, it became the easiest method to acquire property by lawful means. Thus security within the law was established, and one step toward honest actions taken'" (ibid).

20. Thomas, 57.

21. Ibid., 57–58.

22. Ibid.

23. Professor Prosper elaborates: "You must understand that, under the favorable conditions which now surround the human race, all powers for evil are crippled, while those for good are given every possible opportunity for development. The 'survival of the fittest,' which was a new by-word in your day, is now a gospel. . . . The energies that your inventors too often wasted in profitless hide-and-seek with the powers of nature are now directed toward perfecting instruments of every kind for enriching human lives" (ibid., 138). For a useful thematic discussion of the common tendency to eliminate political struggle or conflict within the late-nineteenth-century utopian romance, see Robert Elliott's chapter "The Aesthetics of Utopia" (102–28).

24. Chris Ferns remarks that "by the close of the nineteenth century . . . the notion of progress—of history as a process of continuous advance, with change as a norm, rather than the exception—had become commonplace" (68).

25. The book was first published in installments in *Cosmopolitan* from 1892 to 1893.

26. Howells, *A Traveler from Altruria*, 17, 14–15.

27. Ibid., 279.

28. Another example: In Richard Michaelis's well-known parodic response to Bellamy, *Looking Further Forward*, the outcast Professor Forest, an avowed enemy of Bellamy's Dr. Leete, offers Julian West this quasi-botanical refutation of the "communistic" social arrangements advocated in *Looking Backward*: "Is the leading principle in creation equality or variety? You find sometimes similitude but never conformity. Botanists have carefully compared thousands of leaves, which looked exactly alike at the first glance, but which after close examination were found to possess striking dissimilarities. Inequality is the law of nature and the attempt to establish equality is therefore unnatural and absurd" (30). Here, quite regardless of the author's fervent opposition to Bellamy's version of evolution, the language of biological research is again made the imprimatur of proper sociopolitical speculation, shortcutting the considerable number of logical steps that would actually be required to connect them in a full-scale evolutionary model. Even the tableau of "botanists" arriving at comprehensive theory by comparing "thousands of leaves" is directly traceable to Darwin's and others' plethoric studies of plant and animal traits, which are conventionally referenced as the empirical groundwork for the more grandiose theoretical syntheses of evolutionists.

29. However, Howells's odd ambivalence toward the romance is evident throughout his crucial theorizations of realism, not to mention his turn back to utopian romance itself following the canonically "realistic" *A Hazard of New Fortunes* in 1890. For a useful discussion, see Kaplan, 15–25, 63–64.

30. Satterlee, 44.

31. While Darwinism is rarely quite this jumbled within utopian fiction, it is nonetheless often self-contradictory, since it represents an overly convenient dogma confronted by an inconveniently disparate medley of sociopolitical theories. Thus, even in the work of utopian or dystopian writers who are far more suspicious of science and evolution than Bellamy, Thomas, or Satterlee, for instance in William Morris's *News from Nowhere*, we see similar underlying "conditions" and "energies" taken entirely for granted on the register of sociopolitical *explanation*, a direct extension of "Darwinistic" theory and its Newtonian foundation. In the following passage, the terms of an enlightened, rationalized political evolution replace, in a more clear-sighted future, the obsolete terms of a conflict-based, and therefore obscure and irrational, sociopsychology:

> "I see," said I; "you mean that you have no 'criminal' classes."
>
> "How could we have them," said he, "since there is no rich class to breed enemies against the state by means of the injustice of the state?"
>
> Said I: "I thought that I understood from something that fell from you a little while ago that you had abolished civil law. Is that so, literally?"
>
> "It abolished itself, my friend," said he. . . . "[P]rivate property being abolished, all the laws and all the legal 'crimes' which it had manufactured of course came to an end." (Morris, 69–70)

Morris, despite his preference for handwork and craft over technology and science, nevertheless explicates political history as a transparently mappable play of mechanical forces, a "breed[ing]" process in which social institutions rise or die through the logical or statistical pressure of selection. Likewise, in Winnifred Harper Cooley's 1902 feminist utopia, "A Dream of the Twenty-First Century," a future "instructress" describes "what evolution has done" (207), speaking of "the tendency toward having fewer children" (209), of a "rational religion" in which "simplicity and sense at last conquered" (210) and that occasions the "logical working out of civilization" (210), of women's universal suffrage as "the 'influence' necessary to effect transformations" (209), and in general of a population "bending every force toward the serious business of making life worth living" (211).

32. Kidd, 28. Work such as Kidd's contains the unsurprising heavy dose of ethnocentrism, racism, and sexism that social Darwinism is typically invoked to bolster, along with the usual provisos that "these are the first stern facts of human life" and result from "deep-seated physiological causes, the operation of which we must always remain powerless to escape"—and, moreover, that "[i]t is worse than useless to obscure or to ignore them, as is done in a great part of the social literature of the time" (62).

33. Ibid., 224.

34. Brian Stableford observes, following Frank Manuel, a "shifting of Utopian fantasies into the euchronian mode," which "is correlated with a growing consciousness of the connection between a society's technological resources and its social organization" (*Scientific Romance*, 23–24). Where Manuel sees the "euchronian" shift occurring much earlier than I do, around the end of the eighteenth century, Stableford's examples indicate that he perceives the shift after about the 1870s. Nonetheless, for reasons that will be apparent in this chapter, I pointedly diverge from Stableford's conclusion, concerning the period through the 1880s, that the marketplace in both England and America "afforded little space for evolutionist speculations" (29). See also Manuel, 71.

35. Richard Gerber remarks that "employing such means would have meant a corresponding loss in verisimilitude and concreteness" (99). Similarly, Jean Pfaelzer comments that "in most nineteenth-century utopias and dystopias, applied science, even more than idealized politics, delineates the possibilities of the future" (19). Chris Ferns throws further light on the points made by Gerber and Pfaelzer:

> [I]t is clear that the writer of utopian fiction at the close of the nineteenth century (which sees the first major flowering of utopian fiction since the renaissance) is faced with a radically different relationship between text and context. That dramatic changes in the nature of society are likely to dictate changes in the nature of the alternatives envisaged it goes without saying. . . . What also changes, however, is the nature of the *connection* between the writer's society and the utopian alternatives proposed. . . .
> . . . [I]mperfect reality fills all the territory on the map where utopian perfection could once be imagined. Forced to look elsewhere for a locale, writers of utopias turn increasingly to the space provided by the extension of temporal, rather than geographical horizons. (67–69)

36. Darko Suvin also observes a "shift of SF from space into future time" in the late nineteenth century, albeit a somewhat broader one, which he explains with the suggestion that "setting the tale in the future immediately dispensed with any need for empirical plausibility" (72). I am, of course, arguing for a history that, while not strictly incompatible with Suvin's, indicates a different causal trend within the narrower domain of utopian fiction, in which the "evolutionary setting" is a catalyst precisely for "empirical plausibility." Interestingly, Suvin attributes the shift toward temporality to a "system of mediations . . . [through which] time becomes finally the equivalent of money and thus of all things. . . . 'In short, time becomes space,'" and hence "all existential alternatives, for better or worse, shift into . . . a spatialized future. . . . Positivism shunts SF into anticipation" (73; the subquotation is from Lukács's *History and Class Consciousness*). See also Suvin's sixth chapter, in which, in agreement with

Manuel and Stableford, he describes an even broader historical movement in science fiction from space to time, beginning in the eighteenth century.

37. Howells, for instance, suggests such a temporal analogue when his Altrurian speaks of contemporary Boston as identical to "conditions in my country before the Evolution" (*A Traveler from Altruria*, 191). A similar incipient futurism is present in three contemporaneous English utopias: Edward Bulwer-Lytton's *The Coming Race* (1871), Percy Greg's *Across the Zodiac* (1880), and Richard Jeffries's *After London, or, Wild England* (1886). While it remains generally true, as Kenneth Roemer points out, that "even after the publication of *Looking Backward*, there were many utopias, including Howells's Altrurian Romances, that were set on distant islands, continents, and planets or within the earth," yet even within these examples one readily detects a substantial change in the way that "spatial" utopias are described by their authors, and an increasing bias toward a temporally grounded link between present society and utopia; see Roemer, 82.

38. Pfaelzer, 13–19.

39. Thomas, viii.

40. Ibid., vi.

41. Ibid., viii.

42. Ibid., 301.

43. Ibid., 6.

44. Ibid., 13.

45. Ibid., 13–14.

46. Ibid., 14 (Thomas's emphasis).

47. Ibid., 300.

48. Rosewater, 16. It may be interesting to note that the word "shampoo" is new to the English language in this period; it may be seen as a medical term, belonging to the scientific field of hygiene, alongside "tonic" or the "cataleptic" state produced by shampooing. The word is derived from a Hindi term for a type of therapeutic head massage.

49. Rosewater, 35, 9–10.

50. Ibid., 17ff.

51. Bleiler, 640.

52. Harben, 13.

53. Some of the other futuristic devices present in Harben are a movie projector, a motorized flying machine, and a technologically controlled climate.

54. Morson, 150–55.

55. H. G. Wells, *A Modern Utopia*, 1.

56. Ibid., 2–4.

57. Beckett, *The Unnameable*, 286–87.

58. Note that the very first line of the main apologue of Wells's *A Modern Utopia*, following the italicized prefatory material, credits Darwin as the turning

point between older and "modern" utopian thought: "The Utopia of a modern dreamer must needs differ in one fundamental aspect from the Nowheres and Utopias men planned before Darwin quickened the thought of the world. . . . [T]he Modern Utopia must be not static but kinetic, must shape not as a permanent state but as a hopeful stage, leading to a long ascent of stages" (5).

59. William Burling offers an interpretation of time travel fiction comparable to the one I have been outlining, although my reading diverges from his in significant ways. Burling distinguishes two chief types of time travel fiction, the "temporal dislocation form and the temporal contrast form" (7)—the former is "an exercise in the paradoxes and vagaries of the presumed physics of time and time travel" (8) and has "no interest in social critique" (11), while the latter is "mainly ideological not metaphysical, i.e. the narrative has only a limited (if any) interest in the 'physics' or the 'philosophical implications' of time travel," being instead focused on social contrast, specifically (usually) the "sharp juxtaposition of present-day capitalism with some non-capitalist mode of production" (14). I would argue, from the vantage of the genealogy I have been presenting, that what Burling identifies as the "temporal contrast" form is not (or barely) time travel at all, but rather utopian/dystopian fiction simpliciter. The key indication of this may be the very list of putative time travel texts Burling offers as examples of the "temporal contrast" form, all of which are explicitly either utopian/dystopian, parodic, or nongeneric versions of science fiction: Wells's *The Time Machine*, Marge Piercy's *Woman on the Edge of Time*, Joanna Russ's *The Female Man*, John Kessel's *Corrupting Dr. Nice*, and Ursula Le Guin's *The Dispossessed*. Thus my main disagreement with Burling's useful and cogent analysis is his initial gesture of identifying these works as "time travel narratives" at all, or, more precisely, of contrasting them symmetrically with the "temporal dislocation form."

HISTORICAL INTERVAL I: THE FIRST TIME TRAVEL STORY

1. The protagonist's name is Dr. Moses Nebogipfel in the earliest versions of "The Chronic Argonauts," and Bayliss in one later draft.

2. Mackaye, 1.

3. A good part of *The Panchronicon* consists of exchanges designed to highlight the comic potential of this threesome:

> Droop settled forward with elbows on his knees and brought his finger-tips carefully together. He found this action amazingly promotive of verbal accuracy.
>
> "Well, Cousin Rebecca," he began, slowly, "I'm lookin' fer a partner."
> He paused, considering how to proceed.
>
> The spinster let her hands drop in speechless wonder. The audacity of the man! He—to her—a proposal! At her age! From him!

Fortunately the next few words disclosed her error, and she blushed for it as she lifted her work again, turning nearer the window as if for better light.

"Yes," Droop proceeded, "I've a little business plan, an' it needs capital an' a partner." (9–10)

4. Mackaye, 14.

5. The "International Meridian Conference" was held in October 1884 in Washington, D.C., standardizing the twenty-four-hour day and establishing a world-time scheme with the prime meridian at Greenwich, England. In 1900, one could use a device called a "universal time finder," a circular dial configured like a two-dimensional projection of the Earth, looking down on the north pole, to find the time at any longitude. Spinning this dial backward or forward would have been very much like "cutting meridians." See Lippincott, 148.

6. Mackaye, 16–17.

2. RELATIVITY, PSYCHOLOGY, PARADOX: WERTENBAKER TO HEINLEIN (1923–1941)

1. Einstein, "Relativity Principle," 310.

2. Einstein, "Influence of Gravitation," 379; alternatively, see Einstein, *The Principle of Relativity*, 99.

3. Einstein, "Influence of Gravitation," 387; *The Principle of Relativity*, 108.

4. Eddington, *Space, Time, and Gravitation*, 113.

5. There was ongoing controversy, not really apropos of my own discussion, over both the accuracy of these measurements and the degree to which they confirmed Einstein's prediction.

6. Dyson, Eddington, and Davidson, 332. They also observed deviations in the spectral lines; see Kevles, 175.

7. Lorentz, 25.

8. *Observer* (London) 9 November 1919; *New York Times* 11 November 1919, 17; *Current Opinion* January 1920, 72.

9. *Chicago Tribune* 3 April 1921.

10. The latter quotation is reported as that of Einstein's host, Lord Haldane (*Manchester Guardian* 4 June 1921).

11. See, for instance, Kevles, 139–54.

12. Luckhurst, 53.

13. See Luckhurst's excellent discussion of the ideological positioning of invention and engineering in American science fiction, especially via Hugo Gernsback's earliest pulp magazines in the late 1910s (50–75).

14. This is Hugo Gernsback's phrase, which predates by some years his own now canonical "science fiction." Luckhurst notes that "Gernsback used

the term 'scientific fiction' in 1923, proposed the contraction 'scientifiction' in 1924, which appeared extensively in his editorials of *Amazing Stories* from 1926, but then coined 'science fiction' in his magazine *Science Wonder Stories* in 1929" (15).

15. Einstein, *Relativity*, 42, 90, 52–53, 59–60; Schlick, 11; Eddington, *Theory of Relativity*, 9; Serviss, 45–46.

16. Ashley, Lowndes, and Bleiler are among the only critics to discuss Wertenbaker's work in any detail; see Ashley, 47–51, 57; Ashley and Lowndes, 67, 82f, 257ff; Bleiler, 810.

17. Here is "scientifiction" as used by Hugo Gernsback:

There is the usual fiction magazine, the love story and the sex-appeal type of magazine, the adventure type, and so on, but a magazine of "Scientifiction" is a pioneer in its field in America.

By "scientifiction" I mean the Jules Verne, H. G. Wells, and Edgar Allan Poe type of story—a charming romance intermingled with scientific fact and prophetic vision. ("New Sort of Magazine," 3)

18. Wertenbaker, "The Man from the Atom," 63. All citations of "The Man from the Atom" are from the 1926 reprint in *Amazing Stories* rather than the less available 1923 original in *Science and Invention*.

19. Ibid.

20. Ibid.

21. Ibid., 64–65.

22. Ibid.

23. Jameson, 126.

24. Wertenbaker, "The Man from the Atom," 62.

25. Ibid., 66.

26. Ibid., 63.

27. Ibid.

28. Freedman, 82. Freedman observes that Wells's *The Time Machine* "devotes some of its energy to celebrating a certain kind of bourgeois individualism that (as the Time Traveller's isolated work habits suggest) had been largely occluded by capitalism's corporate phase" (82).

29. Wertenbaker writes, "I have calculated that the effect of a huge foot covering whole countries would be slight, so equally distributed would the weights be" ("The Man from the Atom," 63).

30. The imagery Wertenbaker uses to spatialize the temporal movement he now describes could have been borrowed either from Wells's *The Time Machine* or from more current popularizations of the Einstein theory that Wertenbaker might have seen in films or newspapers: "I remembered how I had seen the streaks that meant the planets going about the sun. So fast had they revolved that I could not see the circuit that meant but a second to me. And yet each

incredibly swift revolution had been a year! A year on earth, a second to me! . . . The few minutes that meant to me the sun's movement through the ether of what seemed a yard had been centuries to the earth" ("The Man from the Atom," 66).

31. Ibid., 66 (Wertenbaker's ellipses).
32. Wertenbaker, "The Man from the Atom (Sequel)," 141.
33. Ibid., 143.
34. Gernsback's preface to the sequel reads:

> In this instalment [*sic*] we find the hero a prisoner on the unknown planet, the inhabitants of which are very much advanced and far superior to the people of the Earth—in intellect and science. His life among these people is not a happy one. Through the interception of a beautiful young girl, some of the best scientists there evolve a method whereby our hero can return to earth [*sic*]. They figure on the basis of Einstein's theory of the curvature of time—if one goes far enough, he will eventually return to where he started from—or in other words "the world having lived and died will live again and die again." It takes millions of years to complete a cycle, but because of the many times increased speed with which our hero travels, because of his enormous size, they are able to figure his return to a time very nearly corresponding to the year in which he left the Earth. Read this imaginative sequel and see how much he succeeds, and how he likes the Earth after he comes back. (Gernsback, "Preface,"141)

35. Wertenbaker, "The Man from the Atom (Sequel)," 144.
36. Ibid. Gernsback appears to believe this idea, as well, perhaps having borrowed it from Wertenbaker: "[A]ccording to Einstein, time, which is a dimension, curves back on itself, and will come back after a certain cycle" ("The Mystery of Time," 525).
37. Wertenbaker, "The Man from the Atom (Sequel)," 143.
38. Ibid., 144–45. Compare: "'What can I tell you,' she said, 'who know so little myself? I have spoken with my uncle. He could not tell me much that I understood. There is some great secret underlying it, some great explanation, which is always just a few steps beyond my grasp. I seem to see for an instant what it is—then suddenly it is gone'" (144).
39. Ibid.
40. Williamson, "The Meteor Girl."
41. Cummings, "Around the Universe"; Leinster, "Sidewise in Time."
42. Hal K. Wells; Simak, "Hellhounds of the Cosmos."
43. Vincent.
44. Breuer.
45. MacFadyen.
46. Frederick; Binder, "The Time Contractor."

47. Leinster, "Sidewise in Time"; Beaumont; P. Schuyler Miller, "The Sands of Time."

48. Schachner; D. L. James; Williamson, "Hindsight."

49. Olsen, "The Four-Dimensional Roller-Press" and "Four Dimensional Surgery"; Cummings, "The White Invaders"; Leinster, "The Fourth-Dimensional Demonstrator"; Binder, "The Time Entity"; A. R. Long, "Scandal in the Fourth Dimension" (or finally, for Leinster, the *fifth* dimension: "The Fifth Dimension Tube").

50. Harry Bates began *Astounding Stories of Super-Science* in 1930 to compete with fantasy magazines like *Amazing Stories*. Defunct by 1933, it was picked up again by Street and Smith, with F. Orlin Tremaine as editor. Campbell took over in May 1938. Isaac Asimov offers this retrospective comment, indicative of the terms a science fiction writer might have had at his or her disposal in the early 1940s: "By the time I was writing 'Super-Neutron,' in February of 1941, I had heard of uranium fission and had even discussed it in some detail with Campbell. I managed to refer to it in the course of the story as 'the classical uranium fission method for power.' I also spoke of the metal cadmium as a neutron absorber. It wasn't bad for a story that appeared in 1941" ("Postscript," 63).

51. Lacan, *Seminar I*, 14.

52. Wertenbaker, "The Man from the Atom (Sequel)," 144.

53. Ibid., 146.

54. Genette, *Narrative Discourse*, 56.

55. Wertenbaker, "The Man from the Atom (Sequel)," 145.

56. Ibid., 146.

57. The version of the story I cite is a reprint of the original *Astounding* text and appears in the collection *Adventures in Time and Space* (1946), still under the pseudonym Anson MacDonald. A slightly altered version is published in Heinlein's own later short story collections, for instance *The Menace from Earth* (1959).

58. MacDonald [Heinlein], 887.

59. Ibid., 890.

60. These books are, as described by Heinlein: "'The Prince,' by Niccoló Machiavelli. 'Behind the Ballots,' by James Farley. 'Mein Kampf' (unexpurgated) by Adolf Schickelgruber. 'How to Win Friends and Influence People,' by Dale Carnegie. . . . 'Real Estate Broker's Manual,' 'History of Musical Instruments,' and a quarto titled 'Evolution of Dress Styles'" (921). In 1941, a direct mention of Hitler may have been somewhat risqué, if not exactly taboo, although the seemingly flippant gesture of substituting the German name "Schickelgruber" is also typical of Heinlein's glib political humor.

61. Rieder, 32.

62. Indeed, because the future setting of "By His Bootstraps" is so similar to a hypothetical primitive "past," John Rieder's discussion of "the anthropolo-

gist's fantasy" in science fiction provides a useful template for understanding the contrast between Diktor's technological savvy and the pretty but decadent innocents he rules:

> Although we know that these people exist here and now, we also consider them to exist in the past—in fact, to be our own past. The point that bears repeating here about this already familiar ideological formation is that, especially in science fiction, technology is the primary way of representing this confrontation of past and present or of projecting it into a confrontation of present and future. The key element linking colonial ideology to science fiction's fascination with new technology is the new technology's scarcity. . . . [T]he relevance of colonialism to stories about technology shows up in the social relations that form around the technology's uneven distribution. (32)

63. A twist is provided in this type of tale of fundamental narcissism by a smaller minority of time travel stories in which, because of a manipulation of the past, the self ceases to exist, even though it may continue to observe or narrate that very disappearance: "But if [the Time Patrol] should fail, then those records had never existed, the Patrol was never founded, Manse Everard never lived. . . . He pushed the thought off, as he always did when it came to haunt him, and concentrated on his work" (Anderson, *The Shield of Time*, 79); "Mightn't the professor have been wrong, so that, as soon as Padway did anything drastic enough to affect all subsequent history, he would make the birth of Martin Padway in 1908 impossible, and disappear?" (de Camp, *Lest Darkness Fall*, 25); "By that very act the telechronic program to correct history vanished, so did the Temporal Institute and so—alas—did I. Tichy himself, being its Director" (Lem, *The Star Diaries*, xii); "'No,' Harrh said, feeling fear thick about him, like a change in atmosphere. 'I don't know of such a piece. I never had such a thing. Check your memory, Alhir.' 'It was from the ruins of the *First Gate*, don't you understand?' And then Alhir did not exist. Harrh blinked, remembered pouring a cup of tea. But he was sitting in the chair, his breakfast before him" (Cherryh, 47).

64. MacDonald [Heinlein], 899–90 (emphasis in all quotations from this story is MacDonald's [i.e., Heinlein's]).

65. Ibid., 903.

66. Žižek, 66.

67. Ibid., 57–58.

68. MacDonald [Heinlein], 891.

69. Ibid., 902. Note that Heinlein's passing observation about Bob's self-loathing—"Wilson decided he did not like the chap's face"—is repeated in numerous time travel self-encounters, and is perhaps an oblique or nascent symptom of the full psychological weight of such misrecognitions—quite different from, say, looking in a mirror—were they to be fully worked through. A character in Leinster's "Sam, This Is You" finds it "irritating not to be able to

call this joker by name" (82); in Baxter's *The Time Ships*, the protagonist is "angered" by his own self's "casual arrogance" (152). In Philip K. Dick's *Now Wait for Last Year*, a character seeing himself is "amazed . . . [by] the physical unattractiveness of the man. He was too fat and a little old. Unpleasantly gray . . . ; do I really look like that? He asked himself morosely" (182). In William Tenn's "Brooklyn Project," the protagonist encounters "a stocky and rather stout man, with a face ludicrously reddened by rage. . . . [H]e was looking at himself and not relishing the sight" (73). A slightly less common counterpart to overcathected self-loathing is overcathected self-love, more likely to be invoked in New Wave time travel fiction from the 1960s on, perhaps as a result of that literature's more self-conscious approach to the erotic ambivalence of science fiction adventure generally, not to mention the familiar narcissism of its focus on the adventuring male protagonist: "'You'! 'I!' It was himself, swooping godlike out of time to bless his enterprise! This was a sort of exchange of love; he was overcome by emotion at the look and feel of this extension of himself, and could bring out no words" (Aldiss, *An Age*, 122).

70. Žižek, 58.

71. MacDonald [Heinlein], 917.

72. In the original *Astounding* version, Heinlein refers, puzzlingly, to "the pathetic fallacy of DesCartes [*sic*]" (917). In the version revised for his story collections, this phrasing is changed, only slightly less puzzlingly, to "the Cartesian fallacy" ("By His Bootstraps," in *The Menace from Earth*, 67).

73. Heinlein, "Life-Line," 197.

74. Ibid.

75. Ibid., 196.

76. Ibid.

77. Ibid., 197.

78. Ibid., 196, 200.

79. The uncanny correspondence of narrative closure with individual death is of course a venerable literary topic. Its most direct or detailed depiction may be Beckett's *Malone Dies*.

80. Patterson suggests Heinlein might have acquired such a concept from his reading of Ouspensky's *Tertium Organum* (Patterson, 147).

81. MacDonald [Heinlein], 925.

82. Ibid., 932–33.

83. Žižek, 58, partly quoting Lacan's seminar *The Four Fundamental Concepts of Psychoanalysis*.

84. MacDonald [Heinlein], 903. The full passage reads:

At least he thought it was the same scene. Did it differ in any respect? He could not be sure, as he could not recall, word for word, what the conversation had been.

For a complete transcript of the scene that lay dormant in his memory he felt willing to pay twenty-five dollars cash, plus sales tax. (902–3)

85. Ibid., 935.
86. Žižek, 57.
87. Ibid.

HISTORICAL INTERVAL II: THREE PHASES OF TIME TRAVEL / *THE TIME MACHINE*

1. Hutcheon, *A Theory of Parody*, 31. This is part of Hutcheon's description of the "dual ontological status" of "the modern novel," also described by Genette as "hypertextual," or by Scholes as "metafictional." See Hutcheon 30ff.

2. Clute and Nicholls identify paradoxes in some very early writers, for example Anstey and even Mackaye, although these often occur more in the form of hints than of worked-through plot devices. There are a handful of anomalously early outright paradox or "loop" stories in the early 1930s, for instance Frank J. Bridge's "Via the Time Accelerator" (1931).

3. Del Rey, 20. Del Rey further remarks, "This was before Robert Heinlein did his superb 'By His Bootstraps,' which was at least as circuitous as my story, but a lot better."

4. Science fiction critics identify a variety of "looping" paradoxes, among them the predestination paradox, ontological paradox, and bootstrap paradox (ostensibly named for Heinlein's story).

5. Clute and Nicholls, 1225.

6. Critics sometimes identify the dinner-guest narrator as the character Hillyer, whom the Time Traveller says he sees in his lab "for a moment" while returning from the future (H. G. Wells, *The Time Machine*, 166).

7. Ibid., 70. Note that Wells places quotation marks around the Time Traveller's entire monologue in *The Time Machine*. For the sake of convenience, in all instances where my citations are of the Time Traveller's monologue, I quote it directly rather than as subquotation.

8. Ibid., 101.

9. Ibid., 107. In an earlier version, published anonymously in the *National Observer* in 1894, the Time Traveller makes this assertion: "In the past . . . the evolution has not always been upward. The land animals, including ourselves, zoologists say, are the descendants of almost amphibious mudfish that were hunted out of the seas by the ancestors of the modern sharks." And when one of his guests asks, "[W]hat [in the future] will become of Social Reform," the Time Traveller responds: "Let us leave social reform [*sic*] to the professional philanthropist. . . . I told you a story; I am not prepared to embark upon a political discussion" (H. G. Wells, *National Observer Time Machine*, 173).

10. H. G. Wells, *The Time Machine*, 103.

11. Ibid., 105.

12. Ibid., 119.

13. Ibid., 103.

14. Jameson, 123.

15. H. G. Wells, *The Time Machine*, 163.

16. Luckhurst, 35; D. Smith, 48.

17. Rieder, 84.

18. H. G. Wells, *The Time Machine*, 81, 109, 154.

19. Ibid., 108–9.

20. Luckhurst, 37.

21. H. G. Wells, *The Time Machine*, 53.

22. Four years after publishing *The Time Machine*, Wells revisits the "long sleep" model of utopian macrologue, with what is generally agreed to be at best mixed aesthetic success, in his next major utopian foray, *When the Sleeper Wakes* (1899). The minor "A Story of the Days to Come" appears in between, in 1897.

23. Istvan Csicsery-Ronay Jr. offers this nice alternative description of the Time Traveller's curiously quaint movement through the fourth dimension: "[H]istory is a sort of European railroad carriage, with ages in contiguous compartments, linked by an autonomous corridor called time, in which time travelers can bicycle up and down in their time machines" ("Futuristic Flu," 29). Michael Sherborne agrees that "with its saddle and controls, the time machine resembles one of the most liberating of Victorian inventions, the bicycle—or more accurately, the tricycle, which was a much more stable vehicle than the nineteenth-century bicycle." Sherborne then notes that "Wells had been riding one since his teens and in 1892 had joined the Cyclist Touring Club" (102–3).

24. See H. G. Wells, *The Time Machine*, 24–33.

25. Ibid., 55–56.

26. Ibid., 30.

27. Ibid., 33.

28. See K. Williams, 27–28; Landon, *The Aesthetics of Ambivalence*, xiii–xvi, 76–77.

29. H. G. Wells, *The Time Machine*, 35.

30. Ibid., 40.

31. Ibid., 42.

32. Ibid., 40.

33. Ibid., 39, 169.

34. Ibid., 169, 40.

35. Ibid., 157.

36. Later illustrators of *The Time Machine* tend to depict the machine as a kind of open car or sledge, at times looking like a high-tech Ford Model T, perhaps a literalizing interpretation of the Time Traveller's hypothesis that

one travels along the fourth dimension exactly as one travels along the other three.

37. H. G. Wells, *The Time Machine*, 58.

38. Ibid., 54–55.

39. Ibid., 165. Keith Williams discusses the discovery of such reversal in the earliest film technologies, in turn relating it to Wells:

> Though the Lumières' first programmes consisted of simple actualities of under one minute such as *Démolition d'un mur*, they quickly discovered (by the serendipity of feeding the reel the wrong way) that cinematic time could run backwards too, so that the wall miraculously recomposed itself from rubble. (Similar effects had already been achieved with Edison's "peepshow" predecessor, the Kinetoscope.) Reversing, as it became known, was quickly exploited for entertainment value. British pioneer Cecil Hepworth recalled delighting audiences by cranking faster, slower, backwards and even stopping his projector to freeze time altogether at particularly grotesque moments. Such novelties find multiple Wellsian parallels. His fantastic invention makes time itself resemble a movie reel, speeded forwards and backwards, or stopped at will. (K. Williams, 2)

40. H. G. Wells, *The Time Machine*, 56–57.

41. G. Stewart, "The 'Videology' of Science Fiction," 160.

42. H. G. Wells, *The Time Machine*, 57–58.

43. Ibid., 157–58.

44. Wells was directly interested in early cinematographic technologies. Shortly following the 1895 publication of *The Time Machine*, he briefly collaborated with the filmmaker Robert Paul in devising a cinematographic "time machine" device that would re-create a journey from past to future using visual and other sensory simulations. Paul applied for a patent in 1895, but the device was never completed. See K. Williams, 27–29; Landon, *The Aesthetics of Ambivalence*, 76, 81–82.

45. H. G. Wells, *The Time Machine*, 35, 36, 46.

3. THE "BIG TIME": MULTIPLE WORLDS, NARRATIVE VIEWPOINT, AND SUPERSPACE

1. Einstein, "Electrodynamics of Moving Bodies," 141. In this passage and the next one I quote I have altered the English translation slightly.

2. Ibid.

3. Born, 3.

4. Ibid.

5. Einstein, *Relativity*, 4.

6. Ibid.

7. Ibid.

8. Ibid.

9. Ibid., 8. In 1908, Minkowski makes the same point in the following way, which will be at least as influential as Einstein's formulation: "Three-dimensional geometry becomes a chapter in four-dimensional physics. You see why I said at the outset that space and time are to fade away into shadows, and that only a world in itself will subsist" (21).

10. Born, 3. Born attributes Einstein's abiding interest in such a principle to Ernst Mach, and alludes to debates over its "positivistic" nature and even its putative tendencies toward "skeptic[ism]." "Nothing," Born suggests, a bit inscrutably or defensively, "was more remote from Einstein's convictions; in later years he emphatically declared himself opposed to positivism" (3).

11. Einstein, *Relativity*, 28.

12. Born calls this effort at consistency "a heuristic principle pointing to weak spots in a traditional theory which has turned out to be empirically unsatisfactory," and he asserts that "it has become the outstanding method of fundamental research in modern physics" (3–4).

13. Luckhurst has a very informative chapter on the "technocultural conjuncture" of the year 1945 (79–91).

14. Frank, 232, 230–31, quoting a letter of Flaubert's.

15. Flaubert, 120 (all ellipses Flaubert's).

16. Frank, 231.

17. Frank, 230, again quoting Flaubert's letter.

18. Frank, 231.

19. Ibid., 225, 232. The posterior nature of this "instant" becomes especially explicit in Frank's discussion of longer modernist works, for instance Joyce's *Ulysses*: "A knowledge of the whole is essential to an understanding of any part; but, unless one is a Dubliner, such knowledge can be obtained only after the book has been read" (235). In Proust, similarly, "the novel the narrator decides to write has just been finished by the reader; and its form is controlled by the method that the narrator has outlined in its concluding pages" (239).

20. Einstein, *Relativity*, 25–33, passim.

21. At other moments in the essay, particularly in his reading of Proust, Frank himself refers to "the transcendent, extra-temporal quality of . . . experiences" that Proust hoped to "translat[e] . . . to the level of esthetic form" (235).

22. J. Miller, 48.

23. Brooks, 25. This notion of narrative as repetition is ultimately grounded in a structuralist understanding of the linguistic signifier as a necessarily "reiterated" speech act, or, more concisely, in what Derrida calls the "primordial structure of repetition" in speech: "speech" itself is "the representation of itself" ("Meaning and Representation," 57).

24. Iser, *The Act of Reading*, 112, 109, 114.

25. My citations refer to the paperback republication of 1952.

26. Williamson, *The Legion of Time*, 44.

27. Ibid., 75.

28. Ibid., 102.

29. Ibid., 24.

30. Williamson's use of the word "travel" of course corresponds to a standard metaphor of narrative "movement" generally. Marie-Laure Ryan, picking up on this conventional language but also recasting it in line with the "possible worlds" model she invokes for narrative analysis, consistently uses "travel" as a description of the reader's activity with respect to fictional possibilities: "While fiction is a mode of travel into textual space, narrative is a travel within the confines of this space" (Ryan, *Possible Worlds*, 5; see also 19, 32, 174).

31. Williamson, *The Legion of Time*, 31. Note that Williamson immediately follows this description with a cinematographic simile, probably echoing Wells's *The Time Machine*, and again anticipating a topic I will revisit in the final two chapters: "The dancing shimmer in that azure mist was oddly disturbing. Sometimes, he thought, he could almost see the outline of some far mountain, the glint of waves, the shapes of trees or buildings—incongruous impressions, queerly flat, two-dimensional, piled one upon another. It was like a movie screen, he thought, upon which the frames were being thrown a thousand times too fast, so that the projected image became a dancing blur" (31).

32. *The Big Time* was originally published in *Galaxy*, March–April, 1958. I cite the 1961 Ace paperback reissue.

33. Leiber, 29.

34. Ibid., 6.

35. Ibid., 6–7.

36. Starting in Chapter 4, in which I discuss some paratextual aspects of Samuel Delany's *Empire Star*, I will examine this relation between paratext and superspace more fully.

37. Leiber, 7.

38. Crowley, 61.

39. Simak, *Highway of Eternity*, 289; *Time Is the Simplest Thing*, 128.

40. Bear, *Eternity*, 87; Moorcock, *The Rituals of Infinity*, 34; Zelazny, 89; Moorcock, "Escape from Evening," 50; Silverberg, *Across a Billion Years*, 7; Cummings, *The Exile of Time*, 67.

41. Brunner, 151.

42. Bear, *Heads*, 135; Brunner, 151, 153, 12.

43. De Camp, *Lest Darkness Fall*, 93; Heath, 42; Busby, *Islands of Tomorrow*, 244.

44. Ball, 136 (Ball's emphasis).

45. Ibid., 149 (Ball's emphasis).

46. "The End of Eternity" was first written for *Galaxy* as a short story but rejected; this original version appears in *The Alternate Asimovs*. The novel-length

revision was published in 1954. In what follows I cite the original short-story version.

47. "The End of Eternity," 178.

48. "The Red Queen's Race," in *The Early Asimov Vol. II*, 245. Compare Poul Anderson's *The Shield of Time*: "The effects of a change propagate across the world at varying speeds" (300).

49. Asimov, "The End of Eternity," 188, 191.

50. Ibid., 189.

51. See Suvin's discussion of "the novum" as an organizing principle of science fiction (63–84 passim).

52. Clarke, 139.

53. Plotinus, II.V.vii (171); Augustine, XI.1 (253).

54. Heinlein, *Assignment in Eternity*, 231.

55. Note that Minkowski explicitly uses the terminology of "points" or "world points" in addition to "lines" and "world lines" (xvii).

56. Rucker, *Master of Space and Time*, 220, 222.

57. Ibid., 222–23.

58. Davies, 487. "Ensemble of universes" is a term used by physicists to refer to the general layout of a multiverse, prior to identifying specific characteristics.

59. Everett's thesis underwent several revisions and title changes both before and after he received his Princeton Ph.D. My citation here is of the original "long version," which Everett and his advisor, John Wheeler, drastically shortened prior to its submission as a doctoral thesis in 1957; at least some of this editing was done in reaction to the negative responses of Niels Bohr and his colleagues, whom Wheeler had consulted concerning Everett's work. The "long version" was finally published sixteen years later (in Dewitt and Graham) under the new title I cite. For a brief discussion of the various revisions of Everett's work, see Byrne, 136–43.

60. DeWitt, 161.

61. Einstein, "Letter to Schrödinger," 35.

62. Ibid., 36.

63. For a useful discussion of Einstein's correspondence with Schrödinger, as well as his unease with the "incompleteness" of quantum theory, see Arthur Fine's chapter 5, entitled "Schrödinger's Cat and Einstein's: The Genesis of a Paradox" (64–85).

64. Rucker, *Master of Space and Time*, 223.

65. Ibid.

66. Note that Leiber himself disagrees with the point I make, but on nebulous grounds couched in more metaphor and wisecracking literary allusion; Greta Forzane says: "The Place is strictly on the Big Time and everybody that should know tells me that time traveling *through* the Big Time is out. It's this

way: the Big Time is a train, and the Little Time is the countryside and we're on the train, unless we go out a Door, and as Gertie Stein might put it, you can't time travel through the time you time travel in when you time travel" (105). Heinlein, on the other hand, does agree (whether or not the grounds are less nebulous): "If higher dimensions were required to 'Hold' a four-dimensional continuum, then the number of dimensions of space and of time were necessarily infinite; each order requires the next higher order to maintain it" (MacDonald [Heinlein], 931).

4. PARADOX AND PARATEXT: PICTURING NARRATIVE THEORY

1. Delany, *Empire Star*, 3. Throughout this chapter, except where otherwise indicated, I cite the Vintage (2002) edition of *Empire Star* published as a "double" with *Babel-17*. Delany has made minor revisions of the text in this edition; for instance, in the final paragraph quoted here, to substitute "Later" for "And later" in the third line from the last, and "maturity" for "manhood" in the last line. Kevin Donaker-Ring has listed online the errata that still afflict the Vintage edition, at least through 2006. Other, more drastic revisions to the original text, some made by others than Delany himself, I discuss later in the chapter.

2. Delany, "About 5,750 Words," 11.

3. Such features become crucial structural elements and convolutions in Delany's later, more famous works, such as *Babel-17*, the *Neveryona* series, and especially *Dhalgren*.

4. Delany, *Empire Star*, 5–6.

5. Genette, *Narrative Discourse*, 35–47.

6. Delany, *Empire Star*, 85.

7. Lem, *The Star Diaries*, 6.

8. Genette writes:

To study the temporal order of a narrative is to compare the order in which events or temporal sections are arranged in the narrative discourse with the order of succession these same events or temporal segments have in the story, to the extent that the story order is explicitly indicated by the narrative itself or inferable from one or another indirect clue . . . . [I]n the classical narrative . . . reconstitution is most often not only possible, because in those texts narrative discourse never inverts the order of events without saying so, but also necessary. (*Narrative Discourse*, 35)

Chatman provides a characteristically pithy gloss: "[T]he classical narrative is a network (or 'enchainment') of kernels offering avenues of choice only one of which is possible" (56; see also 128).

9. Bal, *Narratology*, 6ff.

10. Culler, *The Pursuit of Signs*, 169–70. See also Culler's essay "Fabula and Sjuzhet in the Analysis of Narrative: Some American Discussions."

11. Chatman, 9.

12. Abbott, 40.

13. The translation of *récit* I borrow from Culler, *Pursuit of Signs*, 170. Genette's translator, Jane Lewin, renders *récit* as "narrative," which is of course reasonable but unavoidably confusing in English, given its potential muddling with "discourse" (French *discours*) or "narrating" (French *narration*). "Enunciation" is Culler's translation for *narration* (*Pursuit of Signs*, 189).

14. Culler, *Pursuit of Signs*, 189.

15. Shklovsky writes, in his essay on Laurence Sterne: "The idea of *sjuzhet* is too often confused with the description of events—with what I propose provisionally to call the *fabula*. The *fabula* is, in fact, only material for *sjuzhet* formation" (57; I have altered the translation by reinserting the original "*sjuzhet*" and "*fabula*" for the translator's "plot" and "story").

16. Fludernik, 3.

17. Ibid., 3, 4. Fludernik is of course well aware that differences of "perspective" abound in historical writing: "[H]istorical discourses do not tell a single, unambiguous story since each historian has a particular view of things and tends to emphasize certain aspects of the age and the events being described while omitting others" (3). To some degree, my discussion diverges from Fludernik's by emphasizing the act of narrating over the process of "worldmaking" as the central feature of fictionalizing, thereby rendering the distinction between, say, novelist and historian somewhat less urgent. The touchstone for such debates over the "factual" or "fictional" nature of historiography is Hayden White's theorization of history as narrative process; see *Metahistory*, "Fictions of Factual Representation," and "Value of Narrativity."

18. Shklovsky, 57.

19. H. James, *Theory of Fictions*, 175.

20. I hasten to add that I consider Fludernik's distinction both reasonable and accurate, even given its "radical[ity]"—just perhaps optimistic, or dependent upon a certain understanding of the formal (in contradistinction to psychological or generic) register on which narratives are initially theorized.

21. Todorov, 27; Genette, *Narrative Discourse*, 165; Brooks, 13.

22. Brooks, 13.

23. Todorov, 27 (Todorov's emphasis); Iser, *Fictive and Imaginary*, 281.

24. Chatman, 49.

25. Doležel, 9.

26. Ibid., 2.

27. Ibid., 3; see also Russell, "On Denoting."

28. Searle, 65; Doležel, 11.

29. Doležel, 13, 11. Note that Doležel himself is very far from endorsing either a "one-world semantics" or the kind of "fictional pragmatics of pretense"

that attempts to forestall the former's problematic approach to fictionality. Instead, Doležel quickly moves to shift the theory of fictionality onto a "radically different footing" by "replacing the one-world with a multiple-world framework" (12).

30. Note that, following both Genette in *Narrative Discourse* and Barthes in his well-known essay "The Reality Effect," one can consider "realism" in narratives as attributable to the presence of extraneous detail that, absent any specific role in advancing plot or character development, connotes the representation of "reality" by virtue of its mere, blank presence: "A useless and contingent detail . . . is the medium par excellence of the referential illusion, and therefore of the mimetic effect: it is a *connotator of mimesis*" (Genette, *Narrative Discourse*, 165).

31. Scholes, 60, paraphrasing aspects of the works of Vladimir Propp and Claude Levi-Strauss.

32. See Genette, *Narrative Discourse*, 189–94; Bal, 145–63; Fludernik, 36–39.

33. See Pavel, 73–113 passim; Doležel 22–23; Ronen, 108–43 passim.

34. Hutcheon, *Narcissistic Narrative*, 5. Hutcheon, following Robert Alter, plausibly speaks of a "literary dialectic [that] would suggest a continuum but a gradually evolving one that has logically culminated in metafiction" (5).

35. I have adapted the term from Avital Ronel's *The Telephone Book* (18). In general, such a notion relies on Derrida's understanding of "the relation of the signifier and the signified in Hegelian dialectics" and the specific implication of this relation that "the signifier is sublated (*aufgehoben*) in the process of meaning" (*Dissemination*, 62n47).

36. "*la passion du réel*"; see Badiou, 32, and passim.

37. Gerrig, 3ff.

38. Ibid., 197.

39. Brooks, 106.

40. Freud's "Wolf Man" case study is published under the title "Notes Upon a Case of Obsessional Neurosis" in *Three Case Histories*.

41. Brooks calls it "one of the most daring moments of Freud's thought, and one of his heroic gestures as a writer" (277).

42. Brooks, 276, 273.

43. Delany, *Empire Star*, 2.

44. Ibid.

45. See Genette, 52ff. Genette describes ellipsis and paralipsis (or "lateral ellipsis") as among a series of potential gaps or quasi fillings that literary narratives can utilize in their recounting of events.

46. Delany, *Empire Star*, 67.

47. Ibid., 68–69 (Delany's ellipsis).

48. My term "viewpoint-over-histories" adapts or co-opts Richard Feynman's term "sum over histories," which refers both to a specific method of

calculating the probability paths of particles and to a general interpretation of quantum theory. Both theoretically and methodologically, the "sum over histories" approach considers equally real all possible paths ("histories") a particle might take through spacetime, then integrates them in order to arrive at an "overall probability amplitude" for the path. The method is described in detail in Feynman's *QED*. Some connections of the "sum over histories" model to the "many-worlds" interpretation of quantum theory are discussed by Stephen Hawking in *A Brief History of Time*, and more recently in Hawking and Mlodinow's *The Grand Design*.

49. Delany, *Empire Star*, 89.

50. G. Stewart, *Dear Reader*, passim.

51. Delany, *Empire Star*, 89.

52. Ibid., 32.

53. Ibid., 40.

54. The term "recomplication," also invoked by Delany in a number of his critical pieces, is from James Blish, used originally to describe the writing of Charles Harness, including the latter's time travel novel *The Paradox Men*. The term refers to plots that are simultaneously highly complex and well closed: "[Harness] is . . . one of the best—and I think *the* best—exponents of the extensively recomplicated plot. Unlike Van Vogt's, Harness's packed plots contain no loose ends and work out to rounded wholes which the reader and the student writer can study with confidence" (Blish, "Some Missing Rebuttals," 21).

55. Delany, *Empire Star*, 27.

56. Ibid., 30.

57. Ibid., 44.

58. Ibid.

59. Ibid., 60, 66, 89.

60. The allusion here is to a line in the first of T. S. Eliot's *Four Quartets*, "Burnt Norton."

61. Delany, *Empire Star*, 83–84.

62. Genette's term for the paratext, and indeed the French title (*Seuils*) of his *Paratexts* (see 1–2).

63. Delany, *Empire Star*, 92.

64. The illustrated edition was fully authorized, but the fact that neither the illustrations nor their captions were executed by Delany himself makes their presence in the book at once an enhancement and a potentially overdetermining distraction, with the capacity—as no enthusiast of the covers of science fiction pulp magazines or paperbacks could fail to understand—to subsume or exceed the text itself, supplying all too concretely a more complete tableau that the written word can only strive to evoke through signs and figures. I am aware that the topic of the relation between text and image is sufficiently vast to war-

rant another several books. I refer, for the sake mainly of efficiency, to W. J. T. Mitchell's discussion in *Picture Theory* of the "inextricable weaving together of representation and discourse, the imbrication of visual and verbal experience" (83); especially apropos are chapters 2 through 4. Mitchell cites as a contemporary touchstone for the problem of textual/visual relations Foucault's book on Magritte, *This Is Not a Pipe*, as well as responses to it by Michel de Certeau and Gilles Deleuze. In particular, de Certeau's reaction to Foucault's "optical" writing style echoes what I see in the "visual" pull of the time travel paradox story—the tendency, away from (historical) *fabula* and toward *text*, to which its *sjuzhet* commits it: "On the level of the paragraph or phrase, quotes function in the same way [that is, like images]; each of them is embedded there like a fragment of a mirror, having the value not of a proof but of an astonishment—a sparkle of others. The entire discourse proceeds in this fashion from vision to vision" (de Certeau, 196; also quoted by Mitchell, 71). It may be noted that the one text most apt to include a theoretical discussion of illustrations in books, Genette's otherwise expansive *Paratexts*, does not do so—indeed, Genette throws up his hands at its prospect, saying the topic is "an immense continent" that "exceeds the means of a plain 'literary person'" (406).

65. Delany, *Empire Star* (1983 edition), 124.

66. Ibid., 125–27.

67. Delany is understandably sensitive to such errors. In his essay "5,750 Words," the general theme of which is the importance of individual words and phrases in science fiction and other genre writing, Delany refers to a "printer's error" in "my second published novel" [presumably *Captives of the Flame*] that "practically destroyed the rest of the story" (3).

THEORETICAL INTERVAL: THE PRIMACY OF THE VISUAL IN TIME TRAVEL NARRATIVE

1. Genette calls paratext "the external presentation of a book" (*Paratexts*, 3).

2. Dick, *Divine Invasion*, 60.

3. Ibid., 62.

4. Landon, *The Aesthetics of Ambivalence*, 74, 76.

5. VIEWPOINT-OVER-HISTORIES: NARRATIVE CONSERVATION IN *STAR TREK*

1. deFord, 185.

2. Quoted by Nahin from comments entitled "The Question of Time Traveling," a preface to Henry F. Kirkham's story "The Time Oscillator" (Nahin, 254).

3. Veronica Hollinger, in 1987, goes so far as to state that the grandfather paradox "is never posed as a 'Father Paradox.' It is as if the SF community is evading the Oedipal aspects implicit in its favorite model of temporal paradox" (216n11).

4. The physicist Kip Thorne proposes a "matricide paradox": "If I have a time machine . . . I should be able to use it to go back in time and kill my mother before I was conceived, thereby preventing myself from being born and killing my mother" (508–9). Yet Thorne also skirts the more obvious psychological implications of parricide, suggesting that the real import of his paradox is "the issue of *free will*: Do I, or do I not, as a human being, have the power to determine my own fate?" (509). Certainly "fate" is an oedipal *topos*, but hardly the most familiar one to post-Freudian readers. Incidentally, as early as 1936, C. L. Moore, in her story "Tryst in Time," articulates this uncanny incompatibility of the block universe and human free will: "So you have infinite freedom in all your actions, yet everything you can possibly do is already fixed in time" (135). Obviously, this problem is an old and familiar one, readily borrowed from any number of philosophies.

5. Horwich, 442–44; Vranas, passim. Note that generally neither philosophers nor science fiction writers are likely to speculate directly on the motivation or underlying desire for self-destruction, even though it seems like an obvious and immediate psychological problem within such scenarios. There are exceptions: In Heinlein's *The Door into Summer*, the protagonist has a "horrid thought:. . . what would happen if I sneaked in and cut the throat of my own helpless body?" (261); in Bob Shaw's *The Two-Timers*, a character says to his doubled self: "'At heart, you *want* to kill me. I represent your own guilt. You're in the unique position of being able to pay the supreme penalty—by executing me—yet to live on" (145). In Richard Meredith's *Vestiges of Time*, a protagonist watches with the aloof or sadistic fascination of a mere narrator, but also marking the uncanny mismatch of "I" and "me" in the scene, as numerous duplicates of himself are slaughtered: "The first of *me* died, but more of *me* came on, a dozen, two dozen; and here and there, as the collective *I* rushed forward, the individual *I* took more wounds. One of me was hit in the head, my skull shattered. I died instantly. . . . And I ran forward, a different *me*, a stream of bullets ripping away my left arm, but somehow I still fired with my right until I collapsed in unbearable pain" (89).

6. Anderson, *There Will Be Time*, 60; Anderson, *The Shield of Time*, 261. Compare Busby's hypothesis of an "inertia of change" or a "finite rate of propagation" of causality in *Islands of Tomorrow* (303–4).

7. Varley, 30; Knight, 85; Leiber, 28–29.

8. Moore, "Vintage Season," 361.

9. Pronzini, 116; Moorcock, *An Alien Heat*, 152.

10. Busby, *Getting Home*, 4.

11. Brunner, 151; de Camp, *Lest Darkness Fall*, 174; Willis, 251.

12. Niven, "Rotating Cylinders," 187; Brown, 548.

13. Koontz, 333.

14. Ibid., 167.

15. "Time's Arrow" is a cliffhanger; parts I and II originally aired in June and September 1992, respectively.

16. Presumably *any* time travel in which one is physically present in the past (as opposed to merely observing it either from "afar" or from a metaphysically nonpresent vantage) amounts to "affecting" the past. As Smeenk and Wüthrich assert, therefore, "most philosophers . . . insist that although a time traveller cannot *change* her past, she must still *affect* it" (9).

17. Lewis, "Paradoxes of Time Travel," 75.

18. Ibid.

19. Ibid.

20. Ibid., 77.

21. Ibid.

22. Ibid. Note that Tim travels back originally to 1920, but attempts to shoot his grandfather during the winter of 1921.

23. Ibid., 75–76.

24. As Slater comments, "*Prima facie*, Lewis's solution foists on us a block universe account of time" (2). I don't wish to speculate further on whether a "block," "growing block," or other model is the correct one for resolving time travel paradoxes. A key point, however, is that such models explicitly exclude multiverses.

25. Leinster, "Rogue Star," 21.

26. Heinlein, *The Door into Summer*, 201; Blish, *Galactic Cluster*, 146.

27. Frankowski, 143.

28. A corresponding loop occurs in *12 Monkeys*, Terry Gilliam's extended adaptation of *La jetée*.

29. Vranas, 1.

30. Novikov, *Evolution of the Universe*, 169 (quoted in Toomey, 141).

31. Novikov, *The River of Time*, 254.

32. Hawking, "Chronology Protection Conjecture," 610.

33. Ibid., 603, 610. See also Hawking, "Chronology Protection Conjecture." This is actually not an original argument on Hawking's part; Nahin provides a useful discussion of a number of versions of it (66–72). Note also that more recent statements by Hawking appear somewhat to temper the surety with which he states his conjecture.

34. Thorne, 521.

35. Deutsch, 310. Deutsch asserts, in agreement with Novikov and Hawking, that "in no case does time travel become inconsistent, or impose special constraints on time travellers' behaviour" (311).

36. Gott, 16–17. This is Gott's description of what he calls "the conservative approach," which he himself does not follow.

37. Ibid., 17.

38. Paraphrasing or translating Roland Barthes's terms *noyau* and *catalyse*, Chatman describes major and minor events as "kernels" and "satellites,"

respectively: "In the classical narrative, only major events are part of the chain or armature of contingency. . . . Kernels are narrative moments that give rise to cruxes in the direction taken by events. They are . . . branching points which force a movement into one or two (or more) possible paths. . . . A minor plot event—a *satellite*—is not crucial in this sense. It can be deleted without disturbing the logic of the plot" (Chatman, 53–54).

39. Lewis, "The Paradoxes of Time Travel," 76.

40. Thorne, 521. Here I have quoted Thorne concisely paraphrasing Hawking's chronology protection conjecture.

41. Thorne, cited in Novikov, *The River of Time*, 263. Note that in a multiverse model, which entails more than one possible outcome for an equivocal event, "past-directed time travel would inevitably be a process set in several interacting and interconnected universes" (Deutsch, 310).

42. Gott, 15

43. Aldiss, *No Time Like Tomorrow*, 77.

44. This similarity of capacities has been called by some philosophers the "autonomy principle": "Suppose an agent, T, within a local region of space-time, S, is able to bring about some configuration of matter with feature F. Then in any local region, S', sufficiently similar to S (with respect to the regions' intrinsic characteristics), an agent, T', sufficiently similar to T, must also be able to bring about a configuration of matter with the feature, F" (Chambers, 298 [quoting a suggestion by an anonymous referee of his article]; also see Sider, and Deutsch and Lockwood [71], both of whom Chambers cites). Lewis's more accessible version of this condition is a continuation of his "Tim" anecdote: "Suppose that down the street another sniper, Tom, lurks waiting for another victim, Grandfather's partner. Tom is not a time traveler, but otherwise he is just like Tim: same make of rifle, same murderous intent, same everything. We can even suppose that Tom, like Tim, believes himself to be a time traveler. Someone has gone to a lot of trouble to deceive Tom into thinking so. There's no doubt that Tom can kill his victim; and Tim has everything going for him that Tom does" ("The Paradoxes of Time Travel," 75).

45. Lewis, "The Paradoxes of Time Travel," 75.

46. Ibid., 78, 75.

47. Grey, 66.

48. Freeze, 34.

49. Mitry, 52.

50. Hogan, 74.

51. Disch, 63; Dick, "Schizophrenia," 178.

52. Whoopi Goldberg plays an El-Aurian, a species sufficiently long-lived (500–700 years) that the producers can flirt with the paratextual joke that she may have survived into the show's present from her erstwhile disguise as a contemporary African-American actor.

53. The term "magical negro" is not definitively attributed but commonly credited to Spike Lee, from discussions with film students at Yale and Washington universities in 2001 about films such as *The Legend of Bagger Vance* and *The Green Mile* ("Director Spike Lee Slams 'Same Old' Black Stereotypes").

54. These are all doubled characters from other episodes of *Star Trek*.

55. Miéville, 361.

56. Note that interventions into past events such as the Kennedy assassination often begin as relatively innocuous attempts by historians to record the past, a fantasy of consummate historiography, and only then introduce paradoxical change as an unwitting or malicious breach of the transtemporal protocols that serve as a surrogate for "historical objectivity" within time travel fictions. Among many others, Wilson Tucker's *The Year of the Quiet Sun* explores such a plot, in which historians "propose to record the Kennedy assassination in a way not done before" (72), or "go back to film the Crucifixion" (49). A touchstone for such stories is Ray Bradbury's famous "A Sound of Thunder," in which the inadvertent crushing of a butterfly on a time travel dinosaur hunting expedition irrevocably alters the entire subsequent future.

57. Norton, 22.

58. Data elaborates: "It appears to be a multiphasic temporal convergence in the spacetime continuum . . . , in essence, an eruption of antitime. . . . [Antitime is] a relatively new concept in temporal mechanics. The relationship of antitime to normal time is analogous to the relationship of antimatter to normal matter."

59. On the phenomenological concept of "world," see Heidegger, especially sections 12–14.

60. Meredith, 169.

61. Cavell, 130.

62. Robinson, 203.

63. Dick, *A Scanner Darkly*, 185.

6. OEDIPUS MULTIPLEX, OR, THE SUBJECT AS A TIME TRAVEL FILM: *BACK TO THE FUTURE*

1. The executive producers of the film were Steven Spielberg, Frank Marshall, and Kathleen Kennedy. Robert Zemeckis and Bob Gale wrote the original screenplay in 1980.

2. Quoted in Sigal, 37.

3. The part of Jennifer Parker is played by Elisabeth Shue in the two sequels.

4. Penley, 68.

5. Michael J. Fox is a good choice to carry through such a regressive project, as Zemeckis and Gale must have realized when they hired him to replace the "too dark" Eric Stoltz a few weeks into production. By the mid-1980s, Fox

was already a fully marketable commodity by way of his role on the television show *Family Ties* (NBC, 1982–89). There he played the politically right-wing yet surprisingly (in the show's terms) good-hearted teenager Alex P. Keaton, whose ongoing conflicts with his ex-hippie parents inverted a standard teenage rebellion plot. It is common for critics to view this kind of unoffending regressive twist as basic fodder for mainstream visual culture during the 1980s, a decade that constantly seemed to offer up its *Zeitgeist* as a reaction to that of other historical eras—for instance, as a repudiation of a supposedly overidealistic 1960s or as a recuperation of an ingenuous and equally fantastical 1950s. Zemeckis and Gale, following their Hollywood mentor Steven Spielberg, have never hesitated to paint cultural history in these sorts of broad strokes. All three films in the *Back to the Future* trilogy (1985–90) build upon Fox's credentials as a mainstream American archetype, simultaneously "conservative" (as Alex P. Keaton) and "liberal" (especially, perhaps, as a Canadian émigré), in order to stage an affectionate burlesque of American mainstream ideology and cultural self-fashioning. Indeed, in pop-cultural or pop-political terms, the very presence of Fox already connects the American '80s to the '50s; it is no great step for the film then to link the two eras more directly via the literal mechanism of the time machine.

6. Just before Marty enters the 1955 coffee shop, a car drives by covered with posters and loudspeakers campaigning for "Red Thomas," whose markedly white face contrasts with the face of the older Goldie Wilson advertised earlier in the film (but of course later in history) on a 1985 vehicle.

7. A number of reviewers note this irony. For instance, Tom O'Brien comments that *"Back to the Future* contains several defects, the most subtle being an unintentional racist suggestion about where fifties blacks got their progressive ideas" (440). The most remarkable such "unintentional racist suggestion" is Marty's "invent[ing] rock 'n' roll," as Robert Zemeckis puts it in an interview included on the DVD release of the film. The scene to which Zemeckis refers is the climactic 1955 high school dance, where, as Marty plays "Johnny B. Goode" on stage, one of the musicians, a Marvin Berry, telephones his cousin Chuck from the wings to let him hear the new music. Thus Marty unwittingly becomes the inspiration for Chuck Berry's music and, implicitly, for the subsequent history of rock 'n' roll. The racism implied—or rather taken for granted—by this offensive scene cannot be viewed as merely incidental to the film's basic structure. Co-writer Bob Gale remarks, in the same DVD interview, that this scene was one of the founding ideas for the film's script, and crucial for deciding to make 1955 the film's primal setting even after the film's "future" scenes had been pushed ahead from 1980 to 1985: "Nineteen-fifty-five was important because we wanted Marty to invent rock 'n' roll; that was one of the ideas that we had real early on" (Gale and Zemeckis, "Candid Q and A").

8. See, for instance, L. Sprague de Camp, "Language for Time Travelers."

9. Even before negotiations to place Pepsi in scenes in the film, Zemeckis and Gale already had in mind the use of such tie-ins. In the first screenplay draft, Doc's discovery of time travel is attributed to Marty's accidental spilling of Coke into Doc's laboratory equipment (Gale and Zemeckis, *"Back the Future,* First Draft").

10. Needless to say, this irony and the problems it causes for cultural theory have been around a long time, explicitly at least since Adorno and Horkheimer's *Dialectic of Enlightenment*: "That is the triumph of advertising in the culture industry: the compulsive imitation by consumers of cultural commodities which, at the same time, they recognize as false" (136).

11. Metz, 60.

12. See G. Stewart, *Between Film and Screen*, passim.

13. Kael, 28.

14. Harwood, 93.

15. Andrew Gordon comments: *"Future* distances us from the incest by making both the mother's brazenness and the son's terror laughable.... The desire, guilt, and fear are still attached to the incest taboo, but the audience is comfortable with those feelings because we get a momentary comic bonus from them" (33). Incidentally, Bob Gale remarks, again in the DVD interview: "[M]y favorite line that we wrote in the entire movie is when she says, 'it's like I'm kissing my brother.' And that made it, like, that just solved that whole problem, we were able to actually make that story work. We struggled over that for a really long time" (Gale and Zemeckis, "Candid Q and A").

16. Wood, 165.

17. Harwood, 74.

18. A fine example of time-travel-as-film also appears in the interestingly self-reflexive *Twilight Zone* episode "Once Upon a Time" (1961), in which a janitor played by Buster Keaton travels from 1890 to 1960. The episode begins in the jerky 16-frames-per-second format of silent film with which Keaton is canonically identified, but, at the precise moment of his arrival in 1960, switches to 24-frames-per-second sound film. Both Keaton and the viewer thus toggle between the silence of any extant recording of the 1890s and the literal noise of a contemporary busy street, as the episode directly offers up whatever pop-cultural capital we would need, alongside our recognition of Keaton, to connect the plot to its minimal historiography of cinema.

19. "Invisible Observer" is a cinematographic category adapted from, and by some critics polemically attributed to, V. I. Pudovkin's 1929 book, *Film Technique*, which argues that the camera ought to "correspond to the natural transference of attention of an imaginary observer (who, in the end, is represented by the spectator)" (43).

20. For one thing, we must disagree with Metz that "at the cinema, it is always the other who is on the screen; as for me, I am there to look at him" (51).

Once the object that I see becomes more than a single source of my psychological identification, then I, too, become more than a single subject with a single imaginary—at this point, the "other" on the screen and the "I" in the audience are relatively exchangeable. To put it another way, when the "I" and the "other" are detached from *specific* subjects, as they are within the duplicated scene in *Back to the Future*, does not the ability of Marty to be both "I" and "other" entail the same ability in me, the spectator, provided that I "identify" with this now multiple subject/object?

21. J. Miller, 47.

22. Ibid., 50.

23. Ricoeur, 182.

24. Metz, 51.

25. Ibid.

26. Ibid.

27. Lacan, *Seminar, Book I*, 14.

28. Lacan, "Function and Field of Speech," 86.

29. However, I am partial to an especially clever and claustrophobic instance of filmic self-surveillance that is offered much earlier by the *Twilight Zone* episode "Shadow Play." A death-row inmate—apparently in a kind of schizoid lucid dream—repeatedly relives the lines of plot leading up to his own execution, unable to escape from the abjectly stereotypical dialogue and characters his imagination pilfers straight from Hollywood melodrama, which represents his only familiarity with such events.

30. Landon, *The Aesthetics of Ambivalence*, 74.

31. Moorcock, *The Life and Times of Gerry Cornelius*, 86.

32. Benford, *Sailing Bright Eternity*, 322.

33. Deleuze, 22.

34. Ibid., 23.

35. Ibid.

36. Hume, 153.

37. Deleuze, 29.

38. Bazin, 36.

39. Deleuze, 81. Note that these comments by Deleuze are in reference to Dziga Vertov's "*assemblage*."

40. Ibid.

41. Ibid., 60.

CONCLUSION: THE LAST TIME TRAVEL STORY

1. See Hutcheon, *A Theory of Parody*, 73–74.

2. The story was written in 1958, rejected by *Playboy*, then published in *The Magazine of Fantasy and Science Fiction* in March 1959. Note that in the re-

printed version in the collection *The Unpleasant Profession of Jonathan Hoag*, the story's title is written with an extra dash in front: "—All You Zombies—"; in the table of contents, it has no dashes at all.

3. Christine Jorgensen, who was widely (but erroneously) celebrated as the first recipient of a sex-change operation, had returned to the United States from her surgery and therapy in Denmark in 1952. She is mentioned in Heinlein's story along with Roberta Cowell, an earlier recipient of such surgery in England (140).

4. Heinlein, "All You Zombies—," 151.

5. Ibid., 156.

6. Other time travel writers have constructed versions of Heinlein's fully self-contained *sjuzhet*, most notably David Gerrold in *The Man Who Folded Himself* (1972). Gerrold also published a substantially updated version of this novel in 2003. Another interesting touchstone for self-perpetuating narrative, although considerably less focused on the precise mechanics of the circular plot, is Charles Harness's *The Paradox Men* (1953), originally published as "Flight into Yesterday" in *Startling Stories* in 1949.

7. See also Philip K. Dick's *Counter-Clock World*, a direct precursor to Amis's novel. Rudy Rucker (writing as Rudolf von Bitter Rucker), in the nonfictional work *Geometry, Relativity and the Fourth Dimension*, describes a speculative "antiplanet" in an "antigalaxy," on which "it might be that we would see people living backward in time" (the description closely anticipates the technique of Amis's later *Time's Arrow*): "Everyone was crying. The tears welled out of their handkerchiefs and ran into their eyes. They walked backward up to the grave, where the casket was slowly brought up. The body was taken home and laid in bed. The antiman and his antiwife lived backward together for 30 years" (97).

8. Morrison, 6.

9. Ibid., 4, 10, 14.

10. Ibid., viii–xix.

11. Ibid., 3.

12. Ibid., 36, 21.

13. Ibid., 43.

14. It is also at such a moment that *Beloved* perhaps exceeds the narrative ambition of other excellent nongeneric time travel narratives concerning individual redemption or the reconstruction of social histories, for instance Octavia Butler's *Kindred*, Marge Piercy's *Woman on the Edge of Time*, or John Williams's *Captain Blackman*.

15. Morrison, 248. I have reproduced the unpunctuated spacing of the original text; only the first line contains a period.

16. Shklovsky, 27.

17. Ibid., 30.

18. See Chatman, 36.

19. These diagrams are also reproduced by Shklovsky (56).

20. Sterne, 180. The marbled endpaper, or rather a print reproduction of it, constitutes pages 181–82 of the edition I cite; most editions of *Tristram Shandy* include only a photographic reproduction of the marbled page. A *paraleipomenon* is "a list of things omitted in the body of a work, and appended as a supplement" (563, note XXXVI, #6).

21. Stewart writes: "To mediate is to convey, to bridge, to deliver, to transmit. To demediate is to block, to put signals and signs into remission. To mediate is to cross; to demediate is to double-cross, to sabotage the message function" (*Bookwork*, 102).

22. It is amusing to observe the editor of the Oxford edition of *Tristram Shandy* attempting precisely to "remediate" Sterne's marbled page, reinterpreting it as readable text rather than as an unreadable paratext. He treats the marbling itself as something that *could* be interpreted but that in this instance serves as the merely dialectical negation of an "absolute control" over reading: "The reader could not 'penetrate the moral' of the marbled page since neither Sterne nor the printer could retain absolute control over the marbling process, ensuring that every copy of the page would be unique" (Sterne, 563, note XXXVI, #6).

23. Bester's novel is sometimes cited by its original title, *Tiger, Tiger*.

24. Bester, 235–36.

25. Ibid., 233–34.

26. Farmer, *The Maker of Universes*; Heinlein, *The Number of the Beast*; Pohl, *The Coming of the Quantum Cats*; Lindskold and Zelazny, *Chronomaster*. Parallel universe fiction seems, perhaps predictably, to generate series of novels. In the cases of the novels just mentioned, I cite only the first book in the author's series. See also David Deutsch's *The Fabric of Reality*, in which he argues that experiments in quantum computing may provide empirical evidence of the existence of parallel universes and a multiverse.

27. Multiple-viewpoint construction in film, as in fiction, is obviously not new; it may be observed in such disparate examples as *Rashomon* (1950), *It's a Wonderful Life* (1946), and *Sliding Doors* (1998). Even in these examples, the specific question of how the observation of a specific viewer or viewpoint is entangled with the construction and significance of each "parallel" is at least implicit.

28. In a moment, the zoom will reverse, return to the Chicago picnic, and begin to zoom inward toward the sub-atomic scale. Questions of the transition between physical and unphysical viewpoints are similar in this reverse zoom.

29. See Descartes, 20–22.

30. *The Known Universe* begins with a long view of the Tibetan Himalayas, in lieu of the close shot of individual humans picnicking.

31. Perhaps anticipating such a manipulable interactive narrative device, Susan Stewart refers to what she calls "postliterate genres" that focus a viewer's attention on the explicitly technological reproduction of content, as well as on its capacity to manipulate time: "Once the viewer can manipulate these dimensions, he or she becomes aware of the textuality, the boundaries of the work. Through such manipulations, the viewer can become both reader and authority, in control of the temporality and spatiality of the work" (12).

32. Ellison, "One Life," 42.

33. Ibid.

34. Ellison, "Audio commentary." All subsequent quotations from Ellison are from this commentary.

35. "And one time they, they stomped me hard enough that the third grade teacher, Miss O'Hara, nice woman, came and gathered up my mangled little semicorpse, and took me inside, and they put me in the nurse's room, on the cot, and the window was open. It was, I guess it was spring, and I heard the kids under the window saying, 'dirty Jewish Elephant,' and it puzzled me. I couldn't figure out the connection between 'dirty Jew' and 'elephant.' And then I realized, oh yeah, Jews are supposed to have big noses [laughs]. How charming."

36. "This was after the Depression, and we weren't desperately poor, but we were not very well-to-do, and in those days you had to take pride in the way you were dressed. If you didn't, in a small town, they would say, ugh, look how Serita Ellison and Lou Ellison—Doc Ellison, they called him Doc—send their kid to school, looking like a ragamuffin."

37. Elsewhere in the commentary Ellison offers this anecdote: "But the amazing thing that happened while this scene was being shot at the studio, at the CBS studio which had been the Desilu studio, and the Mary Tyler Moore studio, and Republic: I came to the set to see that shot and I got there late, and I entered the building from the wrong entrance, and I wound up behind the set, and I was standing behind the flat where the little boy had just gone to wash his hands, and it was me, in my age, looking at me as a child looking at my father and me as an adult. And I started to cry, and I couldn't stop crying."

Abbott, H. Porter. "Story, Plot, and Narration." In *The Cambridge Companion to Narrative*. Ed. David Herman. New York: Cambridge UP, 2007. 39–51.

Abish, Walter. *Alphabetical Africa*. New York: New Directions, 1974.

Adorno, Theodor, and Max Horkheimer. *Dialectic of Enlightenment*. Trans. Edmund Jephcott. Stanford, CA: Stanford UP, 2002.

Aldiss, Brian. *An Age*. London: Sphere, 1969.

———. *No Time Like Tomorrow*. New York: Signet, 1963.

"All Good Things . . ." (parts 1 and 2). *Star Trek: The Next Generation*. Season 7, episodes 25–26. First-run syndication. 23 May 1994. Dir. Winrich Kolbe. Television.

Amis, Kingsley. *The Green Man*. Chicago: Academy Chicago, 2005.

Amis, Martin. *Time's Arrow*. New York: Vintage, 1992.

Anderson, Maxwell, and Laurence Stallings. *What Price Glory?* In *Three American Plays*. New York: Harcourt, Brace and Co., 1926.

Anderson, Poul. *The Shield of Time*. New York: Tor, 1991.

———. *There Will Be Time*. New York: Signet, 1973.

Anstey, F. *Tourmalin's Time Checques*. New York: D. Appleton and Co., 1891.

Ashley, Mike. *Time Machines: The Story of the Science Fiction Pulp Magazines from the Beginning to 1950*. Liverpool: Liverpool UP, 2001.

Ashley, Mike, and Robert A. W. Lowndes. *The Gernsback Days*. Holicong, PA: Wildside, 2004.

Asimov, Isaac. *The End of Eternity*. New York: Lancer, 1968 [1955].

———. "The End of Eternity" (original version of the novel *The End of Eternity*). In *The Alternate Asimovs*. Garden City, N. Y.: Doubleday, 1986. 169–249.

———. Postscript to "Super-Neutron." In *The Early Asimov, Book Two*. Greenwich, CT: Fawcett, 1972. 63–67.

———. "The Red Queen's Race." In *The Early Asimov, Book Two*. Greenwich, CT: Fawcett, 1972. 236–59.

Atheling, William, Jr. *See* Blish, James.

Augustine, St. *Confessions.* New York: Penguin, 1961.

Aveling, E. B. *The Darwinian Theory. Its Meaning, Difficulties, Evidence, History.* London, 1884.

"Back There." *The Twilight Zone.* Season 2, episode 49. CBS. 13 January 1961. Dir. David Orrick McDearmon. Television.

Back to the Future. Dir. Robert Zemeckis. Universal, 1985. In *Back to the Future: The Complete Trilogy.*

Back to the Future, Part II. Dir. Robert Zemeckis. Universal, 1989. In *Back to the Future: The Complete Trilogy.*

Back to the Future, Part III. Dir. Robert Zemeckis. Universal, 1990. In *Back to the Future: The Complete Trilogy.*

Back to the Future: The Complete Trilogy. DVD. Hollywood, CA: Universal Studios, 2002.

Badiou, Alain. *The Century.* Trans. Alberto Toscano. Cambridge, UK: Polity, 2007.

Bal, Mieke. *Narratology: Introduction to the Theory of Narrative.* 3rd ed. Toronto: U of Toronto P, 2009.

Ball, Brian. *Timepivot.* New York: Ballantine, 1970.

Barthes, Roland. "The Reality Effect." In *The Rustle of Language.* Trans. Richard Howard. Oxford: Blackwell, 1986. 141–48.

Baxter, Stephen. *Manifold: Origin.* New York: Del Rey, 2003.

———. *Manifold: Space.* New York: Del Rey, 2002.

———. *Manifold: Time.* New York: Del Rey, 2000.

———. *The Time Ships.* New York: Harper Voyager, 1995.

Bazin, André. *What Is Cinema, Volume I.* Trans. Hugh Gray. Berkeley: U of California P, 2005.

Bear, Greg. *City at the End of Time.* New York: Del Rey, 2008.

———. *Eternity.* New York: Tor, 2007.

———. *Heads.* New York: Tor, 1992.

Beaumont, David H. "When the Cycle Met." *Astounding Stories* 18.3 (November 1935): 71–75.

Beckett, Samuel. *Malone Dies* [1951]. Trans. by the author. In *Three Novels.*

———. *Molloy* [1947]. Trans. Patrick Bowles in collaboration with the author. In *Three Novels.*

———. *Three Novels.* New York: Grove, 2009.

———. *The Unnameable* [1953]. Trans. by the author. In *Three Novels.*

Bell, Michael Davitt. *The Problem of American Realism.* Chicago: U of Chicago P, 1993.

Bellamy, Edward. *Looking Backward: 2000 to 1887.* New York: Houghton Mifflin Co., 1926 [1887].

Benford, Gregory. "Exposures." In *In Alien Flesh.* New York: Tor, 1988. 231–47.

———. *Sailing Bright Eternity*. New York: Aspect, 2005.

———. *Timescape*. New York: Pocket Books, 1980.

Bergson, Henri. *Matter and Memory*. Trans. N. M. Paul and W. S. Palmer. New York: Zone, 1988.

Bester, Alfred. *The Stars My Destination*. New York: Vintage, 1996 [1956]. Originally published as *Tiger Tiger*.

Binder, Eando. "The Time Cheaters." *Thrilling Wonder Stories* 15.2 (March 1940): 14–31.

———. "The Time Contractor." *Astounding Stories* 20.4 (December 1937): 143–52.

———. "The Time Entity." *Astounding Stories* 18.2 (October 1936): 73–83.

Bleiler, Everett F. *Science-Fiction: The Early Years*. Kent, OH: Kent State UP, 1990.

Blish, James. *Galactic Cluster*. New York: Signet, 1959.

———. "Weapon Out of Time." *Science Fiction Quarterly* 3 (Spring 1941): 135–40.

Blish, James (writing as William Atheling Jr.). "Some Missing Rebuttals: Winter, 1952–53." In *The Issue at Hand*. Ed. James Blish. Chicago: Advent, 1964. 21–32.

Boeke, Kees. *Cosmic View: The Universe in Forty Jumps*. New York: John Day, 1957.

Borges, Jorge Luis. "The Garden of Forking Paths" [1941]. Trans. Helen Temple and Ruthven Todd. In *Ficciones*. Ed. Anthony Kerrigan. New York: Grove, 1962. 89–101.

Born, Max. *Einstein's Theory of Relativity*. Rev. ed. New York: Dover, 1965 [1920].

Bould, Mark, and Sherryl Vint. *The Routledge Concise History of Science Fiction*. New York: Routledge, 2011.

Bowler, Peter J. *Evolution: The History of an Idea*. Berkeley, CA: U of California P, 2003.

Bradbury, Ray. "A Sound of Thunder" [1952]. In *Golden Apples of the Sun*. New York: Bantam, 1990. 203–14.

Breuer, Miles S. "The Einstein See-Saw." *Astounding Stories* 10.1 (April 1932): 74–89.

Bridge, Frank J. "Via the Time Accelerator." *Amazing Stories* 5.10 (January 1931): 912–24.

Brontë, Emily. *Wuthering Heights*. New York: Norton, 2002.

Brooks, Peter. *Reading for the Plot: Design and Intention in Narrative*. Cambridge, MA: Harvard UP, 1992.

Brown, Fredric. "Experiment." In *Honeymoon in Hell*. New York: Bantam, 1982. 548.

Brunner, John. *Times without Number*. New York: Ace, 1969.

Budd, Louis. "The American Background." In *The Cambridge Companion to American Realism and Naturalism.* Ed. Donald Pizer. Cambridge: Cambridge UP, 1995. 21–46.

Bulwer-Lytton, Edward. *The Coming Race.* London: George Routledge and Sons, 1886 [1871].

Burling, William J. "Reading Time: The Ideology of Time Travel in Science Fiction." *Kronoscope* 6.1 (2006): 5–30.

Busby, F. M. *Getting Home.* New York: Ace, 1987.

———. *Islands of Tomorrow.* New York: Avon, 1994.

Butler, Octavia, *Kindred.* New York: Beacon, 2009.

Butler, Samuel. *Evolution, Old and New.* London: A. C. Fifield, 1911 [1882].

———. *Erewhon, or Over the Range.* Newark, DE: U of Delaware P, 1981 [1872].

The Butterfly Effect. Dir. Eric Bress and J. Mackye Gruber. New Line, 2004.

Byrne, Peter. *The Many Worlds of Hugh Everett III.* New York: Oxford UP, 2010.

Calhoun-French, Diane M. "Time-Travel and Related Phenomena in Contemporary Popular Romance Fiction." In *Romantic Conventions.* Ed. Anne K. Kaler and Rosemary Johnson-Kurek. Bowling Green, OH: Bowling Green State UP, 1999. 100–112.

Cavell, Stanley. *The World Viewed: Reflections on the Ontology of Film.* Enlarged ed. Cambridge, MA: Harvard UP, 1979.

Chambers, Timothy. "Discussion: How Not to Diffuse the Principle Paradox." *Ratio* 12.3 (September 1999): 296–301.

Chatman, Seymour. *Story and Discourse: Narrative Structure in Fiction and Film.* Ithaca, NY: Cornell UP, 1980.

Chavannes, Albert. *In Brighter Climes, or, Life in Socioland.* Knoxville, East Tennessee: Chavannes and Co., 1895.

Cherryh, C. J. *Visible Light.* New York: Daw, 1988.

Chicago Tribune. "Einstein in N. Y.: Even Wife Can't Grasp Theories." April 3, 1921.

Citizen Kane. Dir. Orson Welles. RKO, 1941.

"The City on the Edge of Forever." *Star Trek.* Season 1, episode 28. NBC. 6 April 1967. Dir. Joseph Pevney. Television.

Clarke, Arthur C. *Report on Planet Three, and Other Speculations.* New York: Harper and Row, 1972.

Clute, John, and Peter Nicholls. *The Encyclopedia of Science Fiction.* New York: St. Martin's Griffin, 1995.

Coetzee, J. M. *Waiting for the Barbarians.* New York: Penguin, 1981.

Collins, Wilkie. *The Moonstone.* New York: Penguin, 1994.

———. *The Woman in White.* New York: Penguin, 1999.

Cooley, Winnifred Harper. "A Dream of the Twenty-First Century" [1902]. In *Daring to Dream: Utopian Stories by United States Women: 1836–1919.* Ed. Carol Farley Kessler. Boston: Pandora, 1984. 125–30.

Cortázar, Julio. *Hopscotch*. New York: Pantheon, 1987.

Crichton, Michael. *Timeline*. New York: Ballantine, 1999.

Crowley, John. *Great Work of Time*. New York: Bantam, 1991.

Csicsery-Ronay, Istvan, Jr. "Futuristic Flu." In *Fiction 2000: Cyberpunk and the Future of Narrative*. Ed. George Slusser and Tom Shippey. Athens: U of Georgia P, 1992. 26–45.

———. *The Seven Beauties of Science Fiction*. Middletown, CT: Wesleyan UP, 2008.

Culler, Jonathan. "Fabula and Sjuzhet in the Analysis of Narrative: Some American Discussions." *Poetics Today* 1.3 (1980): 27–37.

———. *The Pursuit of Signs: Semiotics, Literature, Deconstruction*. Ithaca, NY: Cornell UP, 2001.

Cummings, Ray. "Around the Universe." *Science and Invention* 11.3–8 (July–December 1923). Reprinted in *Amazing Stories* 2.7 (Oct. 1927).

———. *The Exile of Time*. New York: Ace, 1964.

———. "The White Invaders." *Astounding Stories* 8.3 (December 1931): 310–65.

Dann, Jack, and Gardner Dozois, eds. *Timegates*. New York: Ace, 1997.

Danto, Arthur. *Narration and Knowledge*. New York: Columbia UP, 2007.

Darwin, Charles. *The Descent of Man*. New York: Penguin, 2004 [1871].

———. *On the Origin of Species by Means of Natural Selection*. New York: Penguin, 2009 [1859].

Davidson, Donald. *Essays on Actions and Events*. New York: Oxford UP, 2001.

Davies, Paul. "Universes Galore: Where Will It All End." In *Universe or Multiverse*. Ed. Bernard Carr. New York: Cambridge UP, 2007. 487–506.

de Camp, L. Sprague, "Language for Time-Travelers." In *The Analog Anthology #1*. Ed. Stanley Schmidt. New York: Davis, 1980. 56–66. Originally published in *Astounding Science-Fiction* 21.5 (July 1938): 63–72.

———. *Lest Darkness Fall*. New York: Pyramid, 1963 [1941].

de Certeau, Michel. *Heterologies: Discourse on the Other*. Trans. Brian Massumi. Minneapolis: U of Minnesota P, 1986.

deFord, Miriam Allen. "The Absolutely Perfect Murder." In *Xenogenesis*. New York: Ballantine, 1969. 181–91.

Déjà Vu. Dir. Tony Scott. Touchstone, 2006.

"Déjà Vu." *The Outer Limits*. Season 5, episode 16. Showtime. 9 July 1999. Dir. Brian Giddens. Television.

Delany, Samuel. "About 5,750 Words." In *The Jewel-Hinged Jaw*. Rev. ed. Middletown, CT: Wesleyan UP, 2009. 1–16.

———. *Dahlgren*. New York: Vintage, 2001.

———. *Empire Star*. In "double" format with Tom Purdom, *The Tree Lord of Imeten*. New York: Ace, 1966 [1965].

———. *Empire Star*. Illustrations by John Jude Palencar. In *Distant Stars*. New York: Ballantine, 1981. 87–213.

———. *Empire Star.* New York: Ballantine, 1983.

———. *Empire Star.* In "double" format with *Babel-17*. New York: Vintage, 2002.

Deleuze, Gilles. *Cinema I: The Movement-Image.* Trans. Hugh Tomlinson and Barbara Habberjam. Minneapolis: U of Minnesota P, 2001 [1983].

del Rey, Lester. Postscript to "The Faithful." In *The Early del Rey, Volume I.* New York: Ballantine, 1976. 16–20.

Dennett, Daniel. *Consciousness Explained.* New York: Back Bay, 1992.

Derrida, Jacques. "Meaning and Representation." In *Speech and Phenomena and Other Essays on Husserl's Theory of Signs.* Trans. David B. Allison. Chicago: Northwestern UP, 1973. 48–59.

———. *Dissemination.* Trans. Barbara Johnson. New York: Continuum, 1981.

Descartes, Réné. *Meditations on First Philosophy.* Trans. John Cottingham. *The Philosophical Writings of Descartes, Volume II.* New York: Cambridge UP, 1984. 1–62.

Deutsch, David. *The Fabric of Reality.* New York: Penguin, 1997.

Deutsch, David, and Michael Lockwood. "The Quantum Physics of Time Travel." *Scientific American* (March 1994): 68–74.

DeWitt, Bryce S. "Quantum Mechanics and Reality." In Dewitt and Graham, *Many-Worlds Interpretation.* 155–65. Originally published in *Physics Today* 23.9 (1970): 30–35.

DeWitt, Bryce S., and Neill Graham, eds. *The Many-Worlds Interpretation of Quantum Mechanics.* Princeton, NJ: Princeton UP, 1973.

Dick, Philip K. *Divine Invasion.* New York: Vintage, 1991 [1981].

———. *Now Wait for Last Year.* New York: Vintage, 1993 [1966].

———. *Time Out of Joint.* New York: Vintage, 2002 [1959].

———. *A Scanner Darkly.* New York: Vintage, 1991 [1977].

———. "Schizophrenia and the Book of Changes" [1965]. In *The Shifting Realities of Philip K. Dick: Selected Literary and Philosophical Writings.* Ed. Lawrence Sutin. New York: Vintage, 1996. 175–82.

———. *Valis.* New York: Bantam, 1981.

"Director Spike Lee slams 'same old' black stereotypes in today's films." *Yale Bulletin and Calendar* 29.21 (2 March 2001).

Disch, Thomas. *Camp Concentration.* New York: Avon, 1971 [1968].

Doctor Who (television series). BBC, 1963–89, 1996, 2005–present.

Doležel, Lubomír. *Heterocosmica: Fiction and Possible Worlds.* Baltimore: Johns Hopkins UP, 1998.

Donaker-Ring, Kevin. "Corrections Remaining to Be Made in *Empire Star* by Samuel R. Delany." http://www.oneringcircus.com/es_errata.html. Accessed 5 May 2011.

Donnelly, Ignatius. *Caesar's Column: A Story of the Twentieth Century.* Chicago: F. J. Shulte and Co., 1890.

Dorfman, Eugene. *The Narreme in the Medieval Romance Epic: An Introduction to Narrative Structures.* Toronto: U of Toronto P, 1969.

Dummett, Michael. *Truth and Other Enigmas*. Cambridge, MA: Harvard UP, 1978.

Dyson, F. W., A. S. Eddington, and C. Davidson. "A Determination of the Deflection of Light by the Sun's Gravitational Field, from Observations Made at the Solar Eclipse of May 29, 1919." *Philosophical Transactions of the Royal Society* 20.570–81 (1 January 1920). 291–333.

Eddington, Arthur Stanley. *Report on the Relativity Theory of Gravitation*. London: Fleetway, 1920.

———. *Space, Time and Gravitation: An Outline of the General Relativity Theory*. London: Cambridge UP, 1921.

———. *The Theory of Relativity and Its Influence on Scientific Thought*. Oxford: Clarendon, 1922.

Einstein, Albert. "On the Electrodynamics of Moving Bodies" [1905]. Trans. Anna Beck. In *The Collected Papers of Albert Einstein, Volume 2 (Supplement: English Translation)*. Princeton, NJ: Princeton UP, 1989. 140–71.

———. "On the Influence of Gravitation on the Propagation of Light." Trans. Anna Beck. In *The Collected Papers of Albert Einstein, Volume 3 (Supplement: English Translation)*. Princeton, NJ: Princeton UP, 1989. 379–87.

———. "Letter to Schrödinger (9 August 1939)." In Schrödinger, Planck, Einstein, Lorentz, *Letters on Wave Mechanics*. Ed. K. Przibram. Trans. Martin J. Klein. New York: Philosophical Library, 1967. 35–36.

———. *The Principle of Relativity*. Trans. W. Perrett and G. B. Jeffery. New York: Dover, 1952 [1923].

———. "On the Relativity Principle and the Conclusions Drawn from It." Trans. Anna Beck. In *The Collected Papers of Albert Einstein, Volume 2 (Supplement: English Translation)*. Princeton, NJ: Princeton UP, 1989. 252–311.

———. *Relativity: The Special and the General Theory*. Trans. Robert Lawson. New York: Three Rivers, 1961 [1916].

Eliot, T. S. *Four Quartets*. Orlando, FL: Mariner, 1971 [1954].

Elliott, Robert. *The Shape of Utopia: Studies in a Literary Genre*. Chicago: U of Chicago P, 1970.

Ellis, Bret Easton. *American Psycho*. New York: Vintage, 1991.

Ellison, Harlan. "Audio commentary for 'One Life, Furnished in Early Poverty.'" *The Twilight Zone, Season 1 (1985–1986)*. DVD. Chatsworth, CA: Image Entertainment, 2004.

———. "One Life, Furnished in Early Poverty." *Orbit 8*. Ed. Damon Knight. New York: G. P. Putnam's Sons, 1970. 27–42.

Enemy of the State. Dir. Tony Scott. Touchstone/Universal, 1998.

Eternal Sunshine of the Spotless Mind. Dir. Michel Gondry. Focus Features, 2004.

Everett, Hugh, III. "'Relative State' Formulation of Quantum Mechanics." In DeWitt and Graham, *Many-Worlds Interpretation*. 141–49.

———. "The Theory of the Universal Wave Function." In DeWitt and Graham, *Many-Worlds Interpretation.* 3–140.

Family Ties (television series). Paramount Television/Ubu, 1982–89.

Faulkner, William. *As I Lay Dying.* New York: Norton, 2009.

Ferns, Chris. *Narrating Utopia: Ideology, Gender, Form in Utopian Literature.* Liverpool: Liverpool UP, 1999.

Feynman, Richard P. *QED: The Strange Theory of Light and Matter.* Princeton, NJ: Princeton UP, 2006.

Fine, Arthur. *The Shaky Game: Einstein, Realism and the Quantum Theory.* Chicago: U of Chicago P, 1996.

Flatley, Guy. "Review of *Back to the Future.*" *Cosmopolitan* (September 1985).

Flaubert, Gustav. *Madame Bovary.* Trans. Eleanor Marx Aveling and Paul de Man. New York: Norton, 2004.

Fludernik, Monika. *An Introduction to Narratology.* Trans. Patricia Häusler-Greenfield and Monika Fludernik. New York: Routledge, 2009.

Foote, Bud. *The Connecticut Yankee in the Twentieth Century: Travel to the Past in Science Fiction.* New York: Greenwood, 1990.

Ford, Ford Madox. *The Good Soldier.* New York: Barnes and Noble, 2005.

Foucault, Michel. *This Is Not a Pipe.* Trans. James Harkness. Berkeley: U of California P, 2008.

Frank, Joseph. "Spatial Form in Modern Literature: An Essay in Two Parts: Part One." *The Sewanee Review* 53.2 (Spring 1945): 221–40.

Frankowski, Leo A. *The Cross-Time Engineer.* New York: Del Rey, 1986.

Frederick, J. George. "The Einstein Express." *Astounding Stories* 15.2 (April 1935): 10–31.

Freedman, Carl. *Critical Theory and Science Fiction.* Middletown, CT: Wesleyan UP, 2000.

Freeze, Wayne. "The History of Temporal Express." In *Time Machines: The Best Time Travel Stories Ever Written.* Ed. Bill Adler. New York: Carroll and Graf, 1998. 31–38.

Frequency. Dir. Gregory Hoblit. New Line, 2000.

Freud, Sigmund. "Notes Upon a Case of Obsessional Neurosis." In *Three Case Histories.* Ed. Philip Rieff. New York: Touchstone, 1996. 1–82.

Futurama (television series). Fox, 1999–2003. Comedy Central, 2008–present.

Gale, Bob, and Robert Zemeckis, "Back to the Future: First Draft." http://www.scifiscripts.com/scripts/back_to_the_future_original_draft.html. Accessed May 10, 2011.

———. "Candid Q and A." *Back to the Future: The Complete Trilogy.* DVD, Disc One. Hollywood, CA: Universal Studios, 2002.

Gaspar, Enrique, y Rimbeau. *El anacronópete.* Barcelona: Biblioteca 'Arte y Letras,' 1887.

Genette, Gérard. *Narrative Discourse.* Trans. Jane E. Lewin. Ithaca, NY: Cornell UP, 1980.

———. *Paratexts: Thresholds of Interpretation*. Trans. Jane E. Lewin. New York: Cambridge UP, 1997.

Gerber, Richard. *Utopian Fantasy*. New York: Routledge and Kegan Paul, 1955.

Gernsback, Hugo. "A New Sort of Magazine." *Amazing Stories* 1.1 (April 1926): 3.

———. "The Mystery of Time." *Amazing Stories* 2.6 (September 1927): 525.

———. "Preface" (to "The Man from the Atom [Sequel]"). *Amazing Stories* 1.2 (May 1926): 141.

———. "The Question of Time Traveling." Prefatory comments to "The Time Oscillator" by Henry F. Kirkham. *Science Wonder Stories* 7 (December 1929): 602.

Gerrig, Richard J. *Experiencing Narrative Worlds: On the Psychological Activities of Reading*. New Haven, CT: Westview, 1998 [1993].

Gerrold, David. *The Man Who Folded Himself*. Rev. ed. Dallas: Benbella, 2003.

———. *The Man Who Folded Himself: The Last Word in Time Machine Novels*. New York: Random House, 1972.

Goodman, Nelson. *Fact, Fiction, and Forecast*. Cambridge, MA: Harvard UP, 1983.

———. *Ways of Worldmaking*. Indianapolis: Hackett, 1978.

Gone with the Wind. Dir. Victor Fleming. Metro Goldwyn-Mayer, 1939.

Gordon, Andrew. "Back to the Future: *Oedipus* as Time Traveler." In *The Worlds of Back to the Future: Critical Essays on the Films*. Ed. Sorcha Ní Fhlainn. Jefferson, NC: McFarland, 2010. 29–48.

Gott, Richard. *Time Travel in Einstein's Universe: The Physical Possibilities of Travel Through Time*. New York: Mariner, 2001.

Greene, Brian. *The Fabric of the Cosmos: Space, Time, and the Texture of Reality*. New York: Anchor, 2006.

Greg, Percy. *Across the Zodiac*. London: Trubner and Co., 1880.

Grey, William. "Troubles with Time Travel." *Philosophy* 74 (1999): 55–70.

Grimwood, Ken. *Replay*. New York: William Morrow, 1998.

Groundhog Day. Dir. Harold Ramis. Columbia, 1993.

Harben, Will N. *The Land of the Changing Sun*. New York: Merriam and Co., 1894.

Harness, Charles. *The Paradox Men*. New York: Crown, 1984 [1953].

———. "Flight into Yesterday." *Startling Stories* 19.2 (May 1949): 9–79.

Harwood, Sarah. *Family Fictions: Representations of the Family in 1980s Hollywood Cinema*. New York: St. Martin's Press, 1997.

Hawking, Stephen. *A Brief History of Time*. Expanded ed. New York: Bantam, 1998.

———. "Chronology Protection: Making the World Safe for Historians." In Stephen Hawking, Kip Thorne, Igor Novikov, Timothy Ferris, and Alan Lightman, *The Future of Spacetime*. New York: Norton, 2002. 87–108.

———. "Chronology Protection Conjecture." *Physical Review D* 46.2 (July 1992) [Received September 1991]: 603–11.

Hawking, Stephen, and Leonard Mlodinow. *The Grand Design.* New York: Bantam, 2010.

Hayles, Katherine N. *The Cosmic Web: Scientific Field Models and Literary Strategies in the 20th Century.* Ithaca, NY: Cornell UP, 1984.

Heath, Peter. *The Mind Brothers.* New York: Prestige, 1967.

Hegel, G. W. F. *Lectures on the Philosophy of World History.* Trans. H. B. Nisbet. New York: Cambridge UP, 1981.

Heidegger, Martin. *Being and Time.* Trans. Joan Stambaugh and Dennis J. Schmdt. Rev. ed. Albany: State U of New York P, 2010.

Heinlein, Robert A. "All You Zombies—." In *The Unpleasant Profession of Jonathan Hoag.* New York: Pyramid, 1961. 143–56. Originally published in *The Magazine of Fantasy and Science Fiction* 16.3 (March 1959): 5–15.

———. *Assignment in Eternity.* New York: Baen, 1981 [1953].

———. "By His Bootstraps." In *The Menace from Earth.* New York: Signet, 1959. 39–88.

———. *The Door into Summer.* New York: Ballantine, 1986 [1956].

———. "Life-Line." In *The Man Who Sold the Moon.* New York: Signet, 1951. 191–212. Originally published in *Astounding Science-Fiction* 23.6 (August 1939): 83–95.

Heinlein, Robert A. (writing as Anson MacDonald). "By His Bootstraps." In *Adventures in Time and Space.* Ed. Raymond J. Healy and J. Francis McComas. New York: Random House, 1946. 889–939. Originally published in *Astounding Science Fiction* 27.2 (October 1941): 9–47.

Heisenberg, Werner. *The Physical Principles of the Quantum Theory.* Trans. Karl Eckart and F. C. Hoyt. New York: Dover, 1930.

———. "The Copenhagen Interpretation of Quantum Theory." In *Physics and Philosophy: The Revolution in Modern Science.* New York: Prometheus, 1999 [1958]. 44–58.

Hofstadter, Richard. *Social Darwinism in American Thought.* Boston: Beacon, 1944.

Hogan, James. *Thrice Upon a Time.* New York: Del Rey, 1980.

Hollinger, Veronica. "Deconstructing the Time Machine." *Science Fiction Studies* 14.2 (July 1987): 201–21.

Horwich, Paul. "On Some Alleged Paradoxes of Time Travel." *The Journal of Philosophy* 72.14 (August 1975): 432–44.

Hot Tub Time Machine. Dir. Steve Pink. MGM, 2010.

Howells, William Dean. *A Hazard of New Fortunes.* New York: New American Library, 1965 [1890].

———. *A Traveler from Altruria.* New York: Harper Bros., 1894.

Huang, Cary, and Michael Huang. *Scale of the Universe: An Interactive Model.* Web page. Primax Studio, 2011. http://primaxstudio.com/stuff/scale_of_universe/. Accessed 2 March 2012.

Hume, David. *An Enquiry Concerning Human Understanding*. Ed. Eric Steinberg. Indianapolis: Hackett, 1977.

Hutcheon, Linda. *Narcissistic Narrative: The Metafictional Paradox*. New York: Routledge, 1991.

———. *A Theory of Parody: The Teachings of Twentieth-Century Art Forms*. Second ed. Chicago: U of Illinois P, 2000.

Il Mare. Dir. Lee Hyun-seung. Sidus, 2000.

Iser, Wolfgang. *The Act of Reading: A Theory of Aesthetic Response*. Baltimore: Johns Hopkins UP, 1980.

———. *The Fictive and the Imaginary: Charting Literary Anthropology*. Baltimore: Johns Hopkins UP, 1993.

———. *The Implied Reader: Patterns of Communication in Prose Fiction from Bunya to Beckett*. Baltimore: Johns Hopkins UP, 1974.

Ishiguro, Kazuo. *When We Were Orphans*. New York: Vintage, 2001.

"It's About Time." *My Little Pony: Friendship Is Magic*. Season 2, episode 20. The Hub (Hasbro). Written by M. A. Larsen. 10 March 2012. Television.

It's a Wonderful Life. Dir. Frank Capra. Liberty, 1946.

Jackson, Shelley. *Patchwork Girl*. Web hypertext. Eastgate, 1995.

James, D. L. "The Cosmo-Trap." *Astounding Stories* 17.2 (April 1936): 98–107.

James, David. *Allegories of Cinema: American Film in the Sixties*. Princeton, NJ: Princeton UP, 1989.

James, Henry. *Theory of Fictions*. Ed. James E. Miller Jr. Lincoln: U of Nebraska P, 1971.

———. *The Turn of the Screw*. In *Collected Stories, Volume 2*. New York: Knopf, 2000. 341–449.

Jameson, Fredric. *Archaeologies of the Future: The Desire Called Utopia and Other Science Fictions*. New York: Verso, 2005.

Jeffries, Richard. *After London, or Wild England*. London: Cassell and Co., 1886.

Kael, Pauline. "The Current Cinema." *The New Yorker* (29 July 1985): 58.

Kaplan, Amy. *The Social Construction of American Realism*. Chicago: U of Chicago P, 1988.

Kessel, John. *Corrupting Dr. Nice*. New York: Tor, 1998.

Kevles, Daniel. *The Physicists: The History of a Scientific Community in Modern America*. Cambridge, MA: Harvard UP, 1995.

Kidd, Benjamin. *Social Evolution*. New York: Macmillan, 1915 [1894].

Knight, Damon. "Arachron." In Dann and Dozois, *Timegates*. 81–100.

Koontz, Dean. *Lightning*. New York: Berkley, 1989.

Krauss, Nicole. *The History of Love*. New York: Norton, 2006.

Lacan, Jacques. "The Function and Field of Speech and Language in Psychoanalysis." In *Écrits: The First Complete Edition in English*. Trans. Bruce Fink. New York: Norton, 2006. 197–268.

———. *Seminar, Book I: Freud's Papers on Technique, 1953–1954.* Trans. John Forrester. New York: Norton, 1998.

———. *Seminar, Book XI: The Four Fundamental Concepts of Psychoanalysis.* Trans. Alan Sheridan. New York: Norton, 1981.

La jetée. Dir. Chris Marker. 1962. In *La jetée/Sans soleil.* DVD. Criterion, 2007.

Landis, Geoffrey A. "Ripples in the Dirac Sea." In *The Best Time Travel Stories of All Time.* Ed. Barry Malzberg. New York: ibooks, 2003.

Landon, Brooks. *The Aesthetics of Ambivalence: Rethinking Science Fiction in the Age of Electronic (Re)Production.* Westport, CT: Greenwood, 1992.

Lanham, Richard. *A Handlist of Rhetorical Terms.* Berkeley: U of California P, 1991.

Last Action Hero. Dir. John McTiernan. Columbia, 1993.

Laumer, Keith. *Assignment in Nowhere.* New York: Berkley, 1968.

Lee, Chang-Rae. *A Gesture Life.* New York: Riverhead, 2000.

Le Guin, Ursula. "Another Story, or, A Fisherman of the Inland Sea." In *A Fisherman of the Inland Sea.* New York: Harper Paperbacks, 1995. 159–207.

———. *The Dispossessed.* New York: Avon, 1975.

Leiber, Fritz. *The Big Time.* London: Ace, 1958. Originally published in *Galaxy* 15.5–6 (March–April 1958).

Leibniz, G. W. *Monadology.* In *Philosophical Texts.* Trans. R. S. Woolhouse and Richard Francks. New York: Oxford UP, 1998.

Leinster, Murray. "The Fifth Dimension Tube." *Astounding Stories* 11.3 (January 1933): 366–415.

———. "The Fourth-Dimensional Demonstrator." *Astounding Stories* 16.4 (December 1935): 100–108.

———. "Rogue Star" [1960]. In *Twists in Time.* New York: Avon, 1960. 7–27.

———. "Sam, This Is You" [1955]. In *Twists in Time.* New York: Avon, 1960. 81–99.

———. "Sidewise in Time." *Astounding Stories* 13.4 (June 1934): 10–47.

Lem, Stanislaw. *The Star Diaries.* Trans. Michael Kandel. New York: Avon, 1976 [1971].

———. "The Time-Travel Story and Related Matters of SF Structuring." *Science Fiction Studies* 1 (Spring 1973): 26–33.

Lewis, David. *Counterfactuals.* Malden, MA: Blackwell, 1973.

———. "The Paradoxes of Time Travel." *Philosophical Papers, Volume II.* New York: Oxford, 1986. 67–80.

———. *On the Plurality of Worlds.* Malden, MA: Blackwell, 1968.

Lindskold, Jane, and Roger Zelazny. *Chronomaster.* Computer game. Capstone Software, 1995.

Lippincott, Kristin. *The Story of Time.* London: Merrell Holberton, 1999.

Long, A. R. "Scandal in the Fourth Dimension." *Astounding Stories* 12.6 (February 1934): 94–101.

Lorentz, H. A. *The Einstein Theory of Relativity: A Concise Statement*. New York: Brentano's, 1920.

Lost (television series). ABC, 2004–10.

Luckhurst, Roger. *Science Fiction*. Malden, MA: Polity, 2005.

Lupoff, Richard A. "12:01 PM." *The Magazine of Fantasy and Science Fiction*. 45.6 (December 1973): 44–58.

MacDonald, Anson. *See* Heinlein, Robert.

MacFadyen, A., Jr. "The Time Decelerator." *Astounding Stories* 17.5 (July 1936): 28–38.

Mackaye, Harold Steele. *The Panchronicon*. New York: Charles Scribner's Sons, 1904.

The Manchester Guardian. "Einstein's London Reception" (4 June 1921).

Manuel, Frank E. "Towards a Psychological History of Utopias." *Utopias and Utopian Thought*. Ed. Frank E. Manuel. Boston: Houghton Mifflin, 1966. 69–98.

Memento. Dir. Christopher Nolan. Summit, 2000.

Meredith, Richard C. *Vestiges of Time*. Chicago: Playboy, 1979.

Metz, Christian. "The Imaginary Signifier." Trans. Ben Brewster. *Screen* 16.2 (1975): 14–76.

Michaelis, Richard. *Looking Further Forward*. New York: Arno, 1971 [1890].

Miéville, China. *Kraken*. New York: Del Rey, 2010.

Miller, J. Hillis. *Reading Narrative*. Norman: Oklahoma UP, 1998.

Miller, P. Schuyler. "As Never Was." *Astounding Science Fiction* 32.5 (January 1944): 31–45.

———. "The Sands of Time." *Astounding Stories* 19.2 (April 1937): 116–38.

Minkowski, Hermann. "Space and Time/Raum und Zeit" [1908]. Trans. D. Lehmkuhl. In *Minkowski Spacetime: A Hundred Years Later*. Ed. Vesselin Petkov. New York: Springer, 2010. xiv–xlii.

Minority Report. Dir. Steven Spielberg. Paramount, 2002.

Mitchell, Edward Page. "The Clock That Went Backward." *New York Sun* (18 September 1881).

Mitchell, W. J. T. *Iconology: Image, Text, Ideology*. Chicago: U of Chicago P, 1986.

———. *Picture Theory*. Chicago: U of Chicago P, 1994.

Mitry, Jean. *Aesthetics and Psychology in the Cinema*. Trans. Christopher King. Bloomington: Indiana UP, 2000 [1963].

Moorcock, Michael. *An Alien Heat*. London: Granada, 1972.

———. *Behold the Man*. New York: Avon, 1968.

———. "Escape from Evening." In *The Time Dweller*. New York: Daw, 1969. 25–52.

———. *The Life and Times of Jerry Cornelius*. London: Quartet, 1976.

———. *The Rituals of Infinity*. New York: Daw, 1978 [1965]. Original U.S. title: *The Wrecks of Time*.

Moore, C. L. "Tryst in Time." In *The Best of C. L. Moore*. Ed. Lester del Rey. New York: Ballantine, 1975. 131–58.

———. "Vintage Season." In *The Best of C. L. Moore*. Ed. Lester del Rey. New York: Ballantine, 1975. 315–64.

Morris, William. *News from Nowhere*. New York: Oxford UP, 2003 [1890].

Morrison, Toni. *Beloved*. New York: Vintage, 2004.

Morson, Gary. *The Boundaries of Genre: Dostoevsky's Diary of a Writer and the Traditions of Literary Utopia*. Austin: U of Texas P, 1981.

"The Most Sensational Event in Physics since Newton." *Current Opinion* 68.1 (January 1920): 72–73.

Nahin, Paul J. *Time Machines: Time Travel in Physics, Metaphysics, and Science Fiction*. Second ed. New York: Springer-Verlag, 1993.

New York Times (11 November 1919).

Nietzsche, Friedrich. *The Gay Science*. Trans. Josefine Nauckhoff. New York: Cambridge UP, 2001.

Niven, Larry. "All the Myriad Ways." In *All the Myriad Ways*. New York: Del Rey, 1971. 1–11.

———. "All the Myriad Ways." In *N-Space*. New York: Tor, 1990. 71–80.

———. "Preface" (to "All the Myriad Ways"). In *N-Space*. New York: Tor, 1990. 68–71.

———. "Rotating Cylinders and the Possibility of Global Causality Violation." In *Convergent Series*. New York: Del Rey, 1979. 183–87.

———. "The Theory and Practice of Time Travel." In *All the Myriad Ways*. New York: Del Rey, 1971. 110–23.

Norton, Andre. *Galactic Derelict*. New York: Ace, 1972.

Novikov, Igor. *Evolution of the Universe*. New York: Cambridge UP, 1983.

———. *The River of Time*. New York: Cambridge UP, 1998.

O'Brien, Tom. "Review of *Back to the Future*." *Commonweal* (9 August 1985).

The Observer [London], (Nov. 9, 1919).

Olsen, Bob. "The Four-Dimensional Roller-Press." *Amazing Stories* 2.3 (June 1927): 302–7.

———. "Four Dimensional Surgery." *Amazing Stories* 2.11 (February 1928): 1078–87.

"Once Upon a Time." *The Twilight Zone*. Season 3, episode 78. CBS. 15 December 1961. Dir. Norman Z. McLeod. Television.

"One Life, Furnished in Early Poverty." *The Twilight Zone* (1980s series). Season 1, episode 11b. CBS. 6 December 1985. Dir. Don Carlos Dunaway. Television.

Ouspensky, P. D. *Tertium Organum: A Key to the Enigmas of the World*. London: Kegan Paul, Trench, Trubner and Co., 1922.

Patterson, William H. *Robert A. Heinlein: In Dialogue with His Century, Volume I: Learning Curve*. New York: Tor, 2010.

Pavel, Thomas G. *Fictional Worlds*. Cambridge, MA: Harvard UP, 1986.

Penley, Constance. "Time Travel, Primal Scene, and the Critical Dystopia." In *Close Encounters: Film, Feminism, and Science Fiction.* Ed. Constance Penley, Elisabeth Lyon, Lynn Spigel, and Janet Bergstrom. Minneapolis: U of Minnesota P, 1991. 63–82.

Pfaelzer, Jean. *The Utopian Novel in America, 1886–1896: The Politics of Form.* Pittsburgh: U of Pittsburgh P, 1984.

Piercy, Marge. *Woman on the Edge of Time.* New York: Fawcett Crest, 1976.

Pizer, Donald. *Realism and Naturalism in Nineteenth-Century American Literature.* Rev. ed. Carbondale: Southern Illinois UP, 1984.

———. *Twentieth Century American Naturalism: An Interpretation.* Carbondale: Southern Illinois UP, 1982.

Planck, Max. "Causality in Nature." In *The Philosophy of Physics.* Trans. W. H. Johnston. New York: Norton, 1936.

Plotinus. *Enneads.* Trans. Stephen MacKenna and B. S. Page. New York: Faber, 1956.

Powers of Ten: A Film Dealing with the Relative Size of Things in the Universe and the Effect of Adding Another Zero. Dir. Charles and Ray Eames. IBM, 1968. In *The Films of Charles and Ray Eames, Volume I.* DVD. Chatsworth, CA: Pyramid, 1990.

"The Principal and the Pauper." *The Simpsons.* Season 9, episode 2. Fox. 28 September 1997. Dir. Stephen Dean Moore. Television.

"Profile in Silver." *The Twilight Zone* (1980s series). Season 1, episode 20a. CBS. 7 March 1986. Dir. John Hancock. Television.

Pronzini, Bill. "On the Nature of Time." In *The Best Time Travel Stories of All Time.* Ed. Sam Moskowitz. New York: ibooks, 2002.

Proust, Marcel. *In Search of Lost Time.* Volumes 1–6. Trans. C. K. Scott Moncrieff, Terence Kilmartin, and D. J. Enright. New York: Modern Library, 2003.

Przibram, K., ed. *Letters on Wave Mechanics: Schrödinger, Planck, Einstein, Lorentz.* Trans. Martin J. Klein. New York: Philosophical Library, 1967.

Pudovkin, V. I. *Film Technique and Film Acting: The Cinematic Writings of V. I. Pudovkin.* Trans. Ivor Montagu. New York: Grove, 1960 [1929].

Putnam, Hilary. *Philosophical Papers, Volume 2: Mind, Language, and Reality.* New York: Cambridge UP, 1975.

Puttenham, George. *The Art of English Poesy.* Ed. Frank Whigham and Wayne A. Rebhorn. Ithaca, NY: Cornell UP, 2007.

Quine, W. V. O. *From a Logical Point of View: Nine Logico-Philosophical Essays.* Second rev. ed. Cambridge, MA: Harvard UP, 1980.

Radway, Janice. *Reading the Romance: Women, Patriarchy and Popular Literature.* Chapel Hill: U of North Carolina P, 1991.

Rashomon. Dir. Akira Kurosawa. Daiei, 1950.

Red Dwarf (television series). BBC, 1988–99.

Ricoeur, Paul. "Narrative Time." In *On Narrative.* Ed. W. J. T. Mitchell. Chicago: U of Chicago P, 1981. 165–86.

Rieder, John. *Colonialism and the Emergence of Science Fiction*. Middletown, CT: Wesleyan UP, 2008.

Robinson, Kim Stanley. *The Memory of Whiteness: A Scientific Romance*. New York: Orb, 1996 [1985].

Rocklynne, Ross. "Time Wants a Skeleton." *Astounding Science-Fiction* 27.4 (June 1941): 9–48.

Roemer, Kenneth M. *Utopian Audiences: How Readers Locate Nowhere*. Amherst: U of Massachusetts P, 2003.

Ronel, Avital. *The Telephone Book*. Lincoln: U of Nebraska P, 1991.

Ronen, Ruth. *Possible Worlds in Narrative Theory*. New York: Cambridge UP, 1994.

Rooney, Charles. *Dreams and Visions: A Study of American Utopias, 1865–1917*. Westport, CT: Greenwood, 1985.

Rosewater, Frank. *'96: A Romance of Utopia*. New York: Arno, 1971.

A Rough Sketch for a Proposed Film Dealing with the Powers of Ten and the Relative Scale of Things in the Universe. Dir. Charles and Ray Eames. In *The Films of Charles and Ray Eames*. DVD. Chatsworth, CA: Pyramid, 1990.

Rucker, Rudolf v. Bitter (Rudy). *Geometry, Relativity and the Fourth Dimension*. New York: Dover, 1977.

Rucker, Rudy. *Master of Space and Time*. New York: Baen, 1984.

Run, Lola, Run. Dir. Tom Tykwer. Westdeutsche Rundfunk, 1998.

Ruse, Michael. *The Darwinian Revolution: Science Red in Tooth and Claw*. Second ed. Chicago: U of Chicago P, 1999.

Russ, Joanna. *The Female Man*. Boston: Beacon, 2000 [1975].

Russell, Bertrand. "On Denoting." In *Logic and Knowledge: Essays 1901–1950*. London: Unwin, 1956. 32–56.

Russett, Cynthia Eagle. *Darwin in America: The Intellectual Response, 1865–1912*. San Francisco: W. H. Freeman and Co., 1976.

Ryan, Marie-Laure. *Avatars of Story*. Minneapolis: U of Minnesota P, 2006.

———. *Possible Worlds, Artificial Intelligence and Narrative Theory*. Bloomington: Indiana UP, 1991.

Satterlee, W. W. *Looking Backward and What I Saw*. New York: Arno, 1971 [1890].

A Scanner Darkly. Dir. Richard Linklater. Warner Independent, 2006.

Schachner, Nat. "Entropy." *Astounding Stories* 16.6 (March 1936): 76–104.

Schlick, Moritz. *Space and Time in Contemporary Physics: An Introduction to the Theory of Relativity and Gravitation*. New York: Oxford UP, 1920.

Scholes, Robert. *Structuralism in Literature: An Introduction*. New Haven, CT: Yale UP, 1974.

Searle, John R. "The Logical Status of Fictional Discourse." In *Expression and Meaning: Studies in the Theory of Speech Acts*. New York: Cambridge UP, 1979. 58–75.

Serviss, Garrett P. *The Einstein Theory of Relativity*. New York: Edwin Miles Fadman, 1923.

"Shadow Play." *The Twilight Zone*. Season 2, episode 26. CBS. 5 May 1961. Dir. John Brahm. Television.

Shaw, Bob. *The Two-Timers*. New York: Ace, 1979 [1968].

Sherborne, Michael. *H. G. Wells: Another Kind of Life*. London: Peter Owen, 2010.

Shklovsky, Victor. "Sterne's *Tristram Shandy*: Stylistic Commentary." In *Russian Formalist Criticism: Four Essays*. Trans. Lee Lemon and Marion Ries. Lincoln: U of Nebraska P, 1965. 25–60.

Sider, T. "A New Grandfather Paradox." *Philosophy and Phenomenological Research* 57 (1997): 139–44.

Sigal, Clancy. "Past Imperfect." *The Listener* (12 December 1985).

Silverberg, Robert. *Across a Billion Years*. New York: Tor, 1969.

———. "Introduction." In *Voyages in Time*. Ed. Silverberg. New York: Tempo, 1970.

———. *Up the Line*. New York: Ballantine, 1969.

Simak, Clifford D. "Hellhounds of the Cosmos." *Astounding Stories* 10.3 (June 1932): 336–53.

———. *Highway of Eternity*. New York: Ballantine, 1986.

———. *Time Is the Simplest Thing*. New York: Crest, 1961.

Slater, Matthew H. "The Necessity of Time Travel (On Pain of Indeterminacy)." *The Monist* 88.3 (2005): 2–8.

Sliding Doors. Dir. Peter Howitt. Paramount, 1998.

Smeenk, Chris, and Christian Wüthrich. "Time Travel and Time Machines." In *The Oxford Handbook of Philosophy of Time*. Ed. Craig Callendar. New York: Oxford UP, 2011. 577–630.

Smith, David C. *H. G. Wells: Desperately Mortal: A Biography*. New Haven, CT: Yale UP, 1986.

Sobchack, Vivian. "Child/Alien/Father: Patriarchal Crisis and Generic Exchange." In *Close Encounters: Film, Feminism, and Science Fiction*. Ed. Constance Penley, Elisabeth Lyon, Lynn Spigel and Janet Bergstrom. Minneapolis: U of Minnesota P, 1991. 3–32.

———. *Screening Space: The American Science Fiction Film*. Second ed. New York: Ungar, 1991.

Source Code. Dir. Duncan Jones. Mark Gordon, 2011.

Stableford, Brian. *Scientific Romance in Britain: 1890–1950*. New York: Palgrave Macmillan, 1985.

———. *The Sociology of Science Fiction*. Rockville, MD: Borgo, 2007.

Star Trek. Dir. J. J. Abrams. Paramount, 2009.

Stephenson, Neal. *Anathem*. New York: Harper Perennial, 2010.

Sterne, Laurence. *The Life and Opinions of Tristram Shandy, Gentleman*. Ed. Ian Campbell Ross. New York: Oxford UP, 1983.

Stewart, Garrett. *Between Film and Screen*. Chicago: U of Chicago P, 1999.

———. *Bookwork: Medium to Object to Concept to Art*. Chicago: U of Chicago P, 2011.

———. *Dear Reader: The Conscripted Audience in Nineteenth-Century British Fiction*. Baltimore: Johns Hopkins UP, 1996.

———. *Framed Time: Toward a Postfilmic Cinema*. Chicago: U of Chicago P, 2007.

———. "The 'Videology' of Science Fiction." In *Shadows of the Magic Lamp: Fantasy and Science Fiction in Film*. Ed. George Slusser and Eric S. Rabkin. Carbondale: Southern Illinois UP, 1985. 159–207.

Stewart, Susan. *On Longing: Narratives of the Miniature, the Gigantic, the Souvenir, the Collection*. Durham, NC: Duke UP, 1993.

"A Stitch In Time." *The Outer Limits*. Season 2, episode 1. Showtime. 14 January 1996. Dir. Mario Azzopardi. Television.

Sumner, William Graham. *Social Darwinism: Selected Essays*. Englewood Cliffs, NJ: Prentice-Hall, 1963.

Suvin, Darko. *Metamorphoses of Science Fiction*. New Haven, CT: Yale UP, 1979.

Tenn, William. "Brooklyn Project." In *Voyages in Time*. Ed. Robert Silverberg. New York: Tempo, 1967. 52–63.

The Terminator. Dir. James Cameron. Orion, 1984.

Terminator 2: Judgment Day. Dir. James Cameron. Tristar, 1991.

Terminator 3: Rise of the Machines. Dir. Jonathan Mostow. Warner Bros., 2003.

Thomas, Chauncey. *The Crystal Button, or, Adventures of Paul Prognosis in the Forty-Ninth Century*. New York: Houghton, Mifflin and Co., 1891.

Thorne, Kip. *Black Holes and Time Warps: Einstein's Outrageous Legacy*. New York: Norton, 1995.

"Time and Punishment" ("Treehouse of Horror V," Segment 2). *The Simpsons*. Season 6, episode 6. Fox. 30 October 1994. Dir. Jim Reardon. Television.

"Time's Arrow, Part I." *Star Trek: The Next Generation*. Season 5, episode 26. First-run syndication. 15 June 1992. Dir. Les Landau. Television.

"Time's Arrow, Part II." *Star Trek: The Next Generation*. Season 6, episode 1. First-run syndication. 21 September 1992. Dir. Les Landau. Television.

"Time to Time." *The Outer Limits*. Season 7, episode 15. Sci Fi. August 11, 2001. Dir. James Head. Television.

Todorov, Tzvetan. *Introduction to Poetics*. Trans. Richard Howard. Minneapolis: U of Minnesota P, 1981.

Toomey, David. *The New Time Travelers: A Journey to the Frontiers of Physics*. New York: Norton, 2007.

Twain, Mark. *A Connecticut Yankee in King Arthur's Court*. Ed. Allison R. Ensor. New York: Norton, 1982 [1889].

12 Monkeys. Dir. Terry Gilliam. Universal, 1995.

Varley, John. "Air Raid." In Dann and Dozois, *Timegates*. 20–35.

Vincent, Harl. "Wanderer of Infinity." *Astounding Stories* 12.1 (March 1933). 100–116.

Vonnegut, Kurt, Jr. *Slaughterhouse-Five*. New York: Dell, 1973 [1969].

Vranas, Peter B. M. "Can I Kill My Younger Self: Time Travel and the Retro-suicide Paradox." *Pacific Philosophical Quarterly* 90 (2009): 520–34.

Walton, Kendall L. *Mimesis as Make-Believe: On the Foundations of the Representational Arts*. Cambridge, MA: Harvard UP, 1990.

Wegner, Phillip E. *Imaginary Communities: Utopia, the Nation, and the Spatial Histories of Modernity*. Berkeley: U of California P, 2002.

Wells, H. G. "The Chronic Argonauts" [1888]. In *The Definitive Time Machine: A Critical Edition of H. G. Wells's Scientific Romance*. Ed. Harry M. Geduld. Indianapolis: Indiana UP, 1987. 135–52.

———. *A Modern Utopia*. Lincoln: U of Nebraska P, 1967 [1905].

———. *The National Observer Time Machine* [1894]. In *The Definitive Time Machine: A Critical Edition of H. G. Wells's Scientific Romance*. Ed. Harry M. Geduld. Indianapolis: Indiana UP, 1987. 154–74.

———. "A Story of the Days to Come" [1897]. In *Tales of Space and Time*. Doubleday and McClure, 1899. 164–324.

———. *The Time Machine: An Invention*. Ed. Leon Stover. Jefferson, NC: McFarland and Co., 1996 [1895].

———. *When the Sleeper Wakes: A Critical Edition of the 1899 London and New York Edition*. Ed. Leon Stover. Jefferson, NC: McFarland, 2000.

Wells, Hal K. "Zehru of Xollar." *Astounding Stories* 9.2 (February 1932): 264–95.

Wertenbaker, G. Peyton. "The Man from the Atom." *Amazing Stories* 1.1 (April 1926): 62–66. Originally published in *Science and Invention* 11.4 (August 1923): 329–30.

———. "The Man from the Atom (Sequel)." *Amazing Stories* 1.2 (May 1926): 140–47.

Wheeler, John A. "Assessment of Everett's 'Relative State' Formulation of Quantum Mechanics." In DeWitt and Graham, *Many-Worlds Interpretation*. 151–53.

White, Hayden. "Fictions of Factual Representation." In *Tropics of Discourse: Essays in Cultural Criticism*. Baltimore: Johns Hopkins UP, 1978. 121–34.

———. *Metahistory: The Historical Imagination in the 19th Century*. Baltimore: Johns Hopkins UP, 1975.

———. "The Question of Narrative in Contemporary Historical Theory." *History and Theory* 23.1 (1984): 1–33.

———. "The Value of Narrativity in the Representation of Reality." In *The Content of the Form: Narrative Discourse and Historical Representation*. Baltimore: Johns Hopkins UP, 1990. 1–25.

Williams, John A. *Captain Blackman*. St. Paul, MN: Coffee House, 2000 [1972].

Williams, Keith. *H. G. Wells, Modernity, and the Movies*. Liverpool: Liverpool UP, 2007.

Williamson, Jack. "Hindsight." *Astounding Stories* 25.3 (May 1940): 98–111.

———. *The Legion of Time*. New York: Pyramid, 1952. Originally published in *Astounding Science-Fiction* 12.3–5 (May–July 1938).

———. "The Meteor Girl." *Astounding Stories* 5.3 (March 1931).

Willis, Connie. "Fire Watch" [1984]. In *Time Machines: The Best Time Travel Stories Ever Written*. Ed. Bill Adler Jr. New York: Carroll and Graf, 2002. 241–78.

Wilson, Robert Charles. *The Chronoliths*. New York: Tor, 2001.

Wittenberg, David. "Oedipus Multiplex, or, The Subject as a Time Travel Film: Two Readings of *Back to the Future*." *Discourse* 28.2–3 (Spring–Fall 2006): 51–77.

———. *Philosophy, Revision, Critique: Rereading Practices in Heidegger, Nietzsche, and Emerson*. Stanford, CA: Stanford UP, 2001.

Wood, Robin. *Hollywood from Vietnam to Reagan*. New York: Columbia UP, 1986.

Woolf, Virginia. *Mrs. Dalloway*. Orlando, FL: Harcourt, 2005.

Word Girl (television series). Scholastic Entertainment/PBS, 2007–present.

Wright, Vincent. *Gadsby: Lipogram Novel*. Vancleave, MS: Ramble House, 2009.

"Yesterday's Enterprise." *Star Trek: The Next Generation*. Season 3, episode 15. First-run syndication. 19 February 1990. Dir. David Carson. Television.

Zelazny, Roger. *Roadmarks*. New York: Del Rey, 1979.

Žižek, Slavoj. *The Sublime Object of Ideology*. New York: Verso, 1989.

Lightning Source UK Ltd.
Milton Keynes UK
UKHW011828171220
375147UK00012B/253

9 780823 249978